Society 5.0

Hitachi-UTokyo Laboratory (H-UTokyo Lab.)

Society 5.0

A People-centric Super-smart Society

 Springer Open

Hitachi-UTokyo Laboratory (H-UTokyo Lab.)
The University of Tokyo
Bunkyo-ku, Tokyo, Japan

Based on a translation from the Japanese language edition: Society 5.0 by Hitachi and The University of Tokyo Joint Research Laboratory Copyright ©Hitachi and The University of Tokyo Joint Research Laboratory, 2018

ISBN 978-981-15-2988-7   ISBN 978-981-15-2989-4   (eBook)
https://doi.org/10.1007/978-981-15-2989-4

This book is an open access publication.

This Springer imprint is published by the registered company Springer Nature Singapore Pte Ltd.
The registered company address is: 152 Beach Road, #21-01/04 Gateway East, Singapore 189721, Singapore

Vision Design: A People-Centric Society Founded on the Merging of Cyberspace and Physical Space

A habitat to support the 100-year life: monitoring robots by our side (Sect. 5.2).
Source: Hitachi Global Center for Social Innovation—Tokyo

A resident-led super-smart society: developing a service to enable greater mobility based on the Person's desire and choices. Source: Hitachi Global Center for Social Innovation—Tokyo

Urban Datarization and Cyberspace-Based Data-Driven Planning

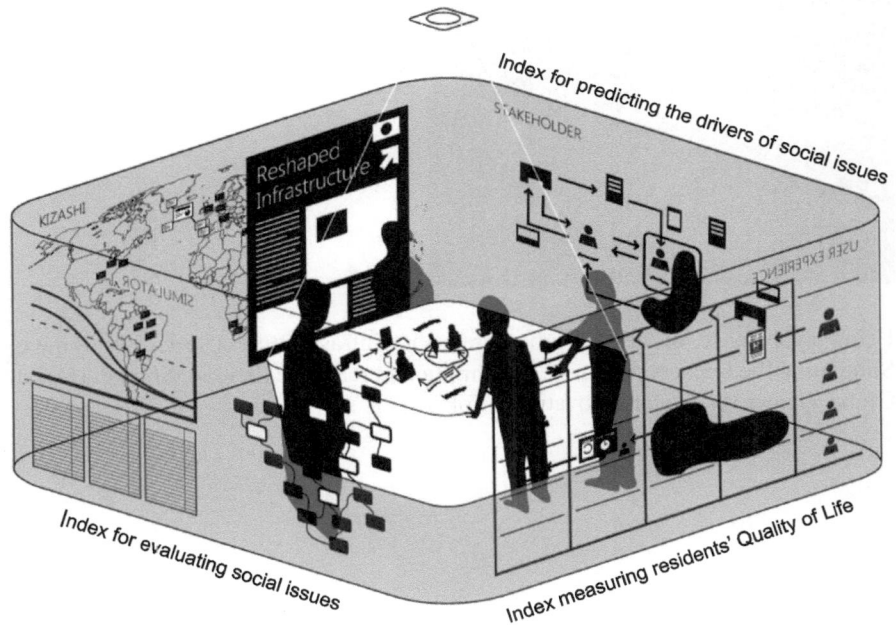

CityScope: using data-driven planning interfaces for town planning (Sect. 5.4). Source: Hitachi Global Center for Social Innovation—Tokyo

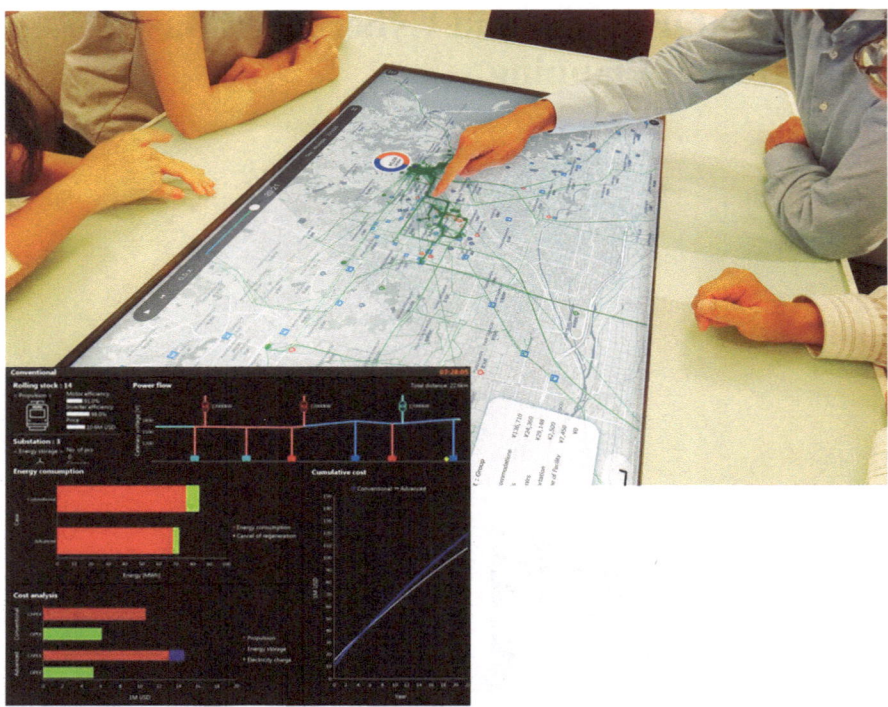

Using cyberspace to design urban transport infrastructure (Sect. 5.4) (above) Simulating the impacts of energy consumption in real time (below). Source: Hitachi Global Center for Social Innovation—Tokyo

Hitachi-UTokyo Laboratory (H-UTokyo Lab.)

Hitachi-UTokyo Laboratory (H-UTokyo Lab.) was founded in 2016 by the University of Tokyo and Hitachi. Rather than following the conventional style of industry-academia partnerships, which focuses on solving specific problems, H-UTokyo Lab has pioneered the industry-academia collaboration model, which pools the strengths of a business and university. Under this model, the Lab creates and communicates a vision for achieving "Society 5.0" and pursues a novel form of research and development intended to address social challenges and make the vision a reality.

Introduction

Big data analytics, artificial intelligence, the Internet of Things—these are just some of the products of research and development that have become regular fixtures of our daily lives. Our private and professional lives are saturated with digital data and information technology through which we develop and share ideas, which in turn generate one new business after another. Just think how our lives have been transformed over the past 10 years, with the rise of the smartphone, new ways of shopping, new ways of working, and the like. If we have changed that much in ten years, then how far have we come over the past 50 years, or even the past 30 years? No one could have imagined the phenomenal change. Digital technology has taken us from an industrial society centered on manufacturing into a society where information is king. Now, we stand at the cusp of a new age. How will we greet this new dawn, and where exactly are we headed?

On January 22, 2016, the Government of Japan released the 5th Science and Technology Basic Plan (Cabinet Office 2016a). The plan proposes the idea of "Society 5.0," a vision of a future society guided by scientific and technological innovation. The intention behind this concept is described as follows: "Through an initiative merging the physical space (real world) and cyberspace by leveraging ICT to its fullest, we are proposing an ideal form of our future society: a 'super-smart society' that will bring wealth to the people. The series of initiatives geared toward realizing this ideal society are now being further deepened and intensively promoted as 'Society 5.0.'"[1] An annotation explains the reasoning behind the term Society 5.0 as follows: "(Society 5.0 is) so called to indicate the new society created by transformations led by scientific and technological innovation, after hunter-gatherer society, agricultural society, industrial society, and information society"(see Fig. 1).

---

[1] See page 13 of *The 5th Science and Technology Basic Plan* (Cabinet Office 2016a). Efforts to address underlying challenges, such as those related to energy, resources, food security, population aging/depopulation, natural disasters, and cyber security, are discussed in sections separate from those concerning Society 5.0. These sections are titled "Sustainable Growth and Self-sustaining Regional Development," "Ensuring Safety and Security for Our Nation and its Citizens and a High-Quality, Prosperous Way of Life," and "Addressing Global Challenges and Contributing to Global Development," and they are found in Chap. 3, which is titled "Addressing Economic and Social Challenges."

	Society 1.0	Society 2.0	Society 3.0	Society 4.0	Society 5.0
Society	Hunter-gatherer	Agrarian	Industrial	Information	Super smart
Productive approach	Capture/Gather	Manufacture	Mechanization	ICT	Merging of cyberspace and physical space
Material	Stone·Soil	Metal	Plastic	Semiconductor	Material 5.0*
Transport	Foot	Ox, horse	Motor car, boat, plane	Multimobility	Autonomous driving
Form of settlement	Nomadic, small settlement	Fortified city	Linear (industrial) city	Network city	Autonomous decentralized city
City ideals	Viability	Defensiveness	Functionality	Profitability	Humanity

Fig. 1 Contextualizing Society 5.0. Categories created by the authors. Source: Produced by authors. *Research conducted by the University of Tokyo's Material Innovation Research Center

In 2016, the government released the "Comprehensive Strategy on Science, Technology and Innovation for 2016" (Cabinet Office 2016b). In the following year, it released the 2017 edition of its comprehensive strategy (Cabinet Office 2017), in which it further described Society 5.0 as follows: "Society 5.0, the vision of future society tow[ard] which the Fifth Basic Plan proposes that we should aspire, will be a human-centered society that, through the high degree of merging between cyberspace and physical space, will be able to balance economic advancement with the resolution of social problems by providing goods and services that granularly address manifold latent needs regardless of locale, age, sex, or language to ensure that all citizens can lead high-quality, lives full of comfort and vitality."(Cabinet Office 2017)

In other words, Society 5.0 is a model to communicate the government's vision of a future society to industry and the general public. This model was the culmination of numerous discussions among experts from various fields. It was also based on research into the history of technology and social development. However, the government literature cited above only provides a brief outline of such scholarly discourse. Without understanding the underlying ideas, one cannot gain a full picture of Society 5.0. What, for example, is cyberspace? What is physical space? What does it mean to merge these two spaces? What does it mean to balance economic advancement with the resolution of social problems? A *human*-centered society—does that not go without saying? Readers would be forgiven for asking such questions. To get the answers, we must understand the thinking and narratives underlying Society 5.0. Hence, this book offers readers a primer on Society 5.0 by discussing the definitions in terms of their implicit meanings and the backdrop from which they emerged.

This book summarizes the findings of the Habitat Innovation project by Hitachi-UTokyo Laboratory (H-UTokyo Lab.). H-UTokyo Lab. was founded in June 2016 following an agreement between the University of Tokyo and Hitachi. Its purpose is to pioneer a new form of industrial-academic partnership known as industry-academia collaboration. Stepping beyond conventional industry-academia partnerships, industry-academia collaboration emphasizes radical and far-reaching inter-institutional coordination as a way of addressing social issues.

This book is primarily authored by members of the H-UTokyo Lab project team as well as by academics from the University of Tokyo. Chapter 1 unpacks the general thinking behind Society 5.0 and lists the relevant nomenclature. Chapter 2 deals with the question of how we can balance what is best for society with what is best for the individual, a question that must be tackled if we are to address social problems under the framework of Society 5.0. The chapter discusses a unique approach to this question: habitat innovation.

Chapter 3 focuses on developments in this century. In particular, it analyzes the rise of the smart city, reviews Japan's efforts to develop the sustainable city, and discusses how these matters relate to Society 5.0.

Chapter 4 discusses urban datarization, an essential requirement for building cyberspace. It also discusses the methods and challenges of integrating different data and systems. Chapter 5 focuses on the work of researchers from the field of engineering. The chapter discusses how such researchers pursue R&D. It also discusses the basic thinking underlying research projects aimed at addressing social problems, including those related to the aging population, the need to go carbon-free, and the need to regenerate rural communities.

Chapter 6 focuses on researchers in the humanities and social sciences. The chapter identifies the key challenges of pursuing a model of society and derives possible approaches to such an end. It also examines what is meant by a people-centric society.

Chapter 7 features a dialogue between Makoto Gonokami, President of the University of Tokyo, and Hiroaki Nakanishi, Chairman of Hitachi. The two leaders discuss the possibilities of Society 5.0 and the direction in which we are headed. Chapter 8 summarizes the challenges we face on the road to Society 5.0 and the prospects for achieving this vision.

We hope that this book will help readers better understand the concept of Society 5.0 and the kind of society it portrays. We also hope that the book will spur discussions between engineers, social scientists, and other experts about the relationship between technology and society, and how this relationship will evolve in the future.

Tokyo, Japan   Atsushi Deguchi
Tokyo, Japan   Osamu Kamimura

References

Cabinet Office (Council for Science, Technology and Innovation) (2016a) The 5th Science and Technology Basic Plan (released on January 22, 2016). https://www8.cao.go.jp/cstp/english/basic/5thbasicplan.pdf. Accessed 4 Jun 2019

Cabinet Office (Council for Science, Technology and Innovation) (2016b) Comprehensive Strategy on Science, Technology and Innovation (STI) for 2016 (released on May 24, 2016). https://www8.cao.go.jp/cstp/sogosenryaku/2016.html. Accessed 4 Jun 2019 https://www8.cao.go.jp/cstp/english/doc/2016stistrategy_summary.pdf (Summarized English version). Accessed 4 Jun 2019

Cabinet Office (Council for Science, Technology and Innovation) (2017) Comprehensive Strategy on Science, Technology and Innovation (STI) for 2017 (released on June 2, 2017), pp. 2. https://www8.cao.go.jp/cstp/english/doc/2017stistrategy_main.pdf. Accessed 4 Jun 2019

Acknowledgments

Many people contributed to this publication. The names of those who contributed are too numerous to list here, but we would particularly like to acknowledge Toshihiko Koseki (former Executive Vice President, The University of Tokyo), Shinobu Yoshimura (Vice President, The University of Tokyo), Norihiro Suzuki (Vice President and Executive Officer, CTO, Hitachi, Ltd.), and Shinji Yamada (General Manager, Center for Exploratory Research, Hitachi, Ltd.) for supporting H-UTokyo Lab on a daily basis and contributing invaluable ideas to this book. Additionally, this book would not have been possible were it not for the support of the University of Tokyo's University Corporate Relations Department, whose members include Takashi Haga and Miho Sugimoto, and the support of the Hitachi R&D Group's Technology Strategy Office, whose members include Mayumi Fukuyama and Tomiko Kinoshita. We also wish to thank Eriko Honda and other members of H-UTokyo Lab's Secretariat for helping to organize the authors' mini symposia. Our thanks also go out to Yoshitaka Shibata and other members of Hitachi's Global Center for Social Innovation—Tokyo for sharing their image data with us. We would like to thank Editage (www.editage.com) for English language editing. We also extend our sincere thanks to everyone else involved in this publication. Finally, we would like to express our deep gratitude to Shuichi Hirai (Editing Division, Nikkei Publishing Inc.) and Mei Hann Lee (Editor, Springer Nature) for giving us the opportunity to publish this book.

Acknowledgments

Contents

Contributors

Atsushi Deguchi Department of Socio-Cultural Environmental Studies, Graduate School of Frontier Sciences, The University of Tokyo, Tokyo, Japan

Hideyuki Matsuoka Center for Exploratory Research, Research & Development Group, Hitachi, Ltd., Tokyo, Japan

Chiaki Hirai Global Center for Social Innovation—Tokyo, Research & Development Group, Hitachi, Ltd., Tokyo, Japan

Osamu Kamimura Industry-Academia-Government Collaboration Department, Technology Strategy Office, Research & Development Group, Hitachi, Ltd., Tokyo, Japan

Taku Nakano Department of Housing and Urban Planning, Building Research Institute, Ibaraki, Japan

Kohei Oshima Department of Urban Engineering, Graduate School of Engineering, The University of Tokyo, Tokyo, Japan

Mitsuharu Tai System Innovation Center, Research & Development Group, Hitachi, Ltd., Tokyo, Japan

Shigeyuki Tani Social Systems Engineering Research Department, System Innovation Center, Research & Development Group, Hitachi, Ltd., Tokyo, Japan

Ryosuke Shibasaki Center for Spatial Information Science, The University of Tokyo, Tokyo, Japan

Satoru Hori Social Systems Engineering Research Department, Center for Technology Innovation—Systems Engineering, Research & Development Group, Hitachi, Ltd., Tokyo, Japan

Shunji Kawamura Security Research Department, Center for Technology Innovation—Systems Engineering, Research & Development Group, Hitachi, Ltd., Tokyo, Japan

Yasunori Akashi Department of Architecture, Graduate School of Engineering, The University of Tokyo, Tokyo, Japan

Eiji Hato Department of Civil Engineering, Graduate School of Engineering, The University of Tokyo, Tokyo, Japan

Junichiro Ohkata Institute of Gerontology, The University of Tokyo, Tokyo, Japan

Shin'ichi Warisawa Department of Human and Engineered Environment Studies, Graduate School of Frontier Sciences, The University of Tokyo, Tokyo, Japan

Shinji Kajitani Department of Interdisciplinary Cultural Studies, Graduate School of Arts and Sciences, The University of Tokyo, Tokyo, Japan

Takahiro Nakajima Institute for Advanced Studies on Asia, The University of Tokyo, Tokyo, Japan

Hiroshi Ohashi Graduate School of Public Policy, The University of Tokyo, Tokyo, Japan

Graduate School of Economics, The University of Tokyo, Tokyo, Japan

Tsutomu Watanabe Graduate School of Economics, The University of Tokyo, Tokyo, Japan

Kaori Karasawa Division of Socio-Cultural Studies, Graduate School of Humanities and Sociology, The University of Tokyo, Tokyo, Japan

Shigetoshi Sameshima Center for Technology Innovation, Research & Development Group, Hitachi, Ltd., Tokyo, Japan

The original version of this book was revised with the Correct Institutional Editor's name as "Hitachi-UTokyo Laboratory (H-UTokyo Lab.)" and also this book was inadvertently published with an incorrect license type 'CC BY 4.0'. The Open Access License has been amended throughout the book to the correct license type 'CC-BY- NC-ND'. The correction to this book is available at https://doi.org/10.1007/978-981-15-2989-4_9

Chapter 1
What Is Society 5.0?

Atsushi Deguchi, Chiaki Hirai, Hideyuki Matsuoka, Taku Nakano, Kohei Oshima, Mitsuharu Tai, and Shigeyuki Tani

Abstract This chapter elaborates on the general thought process behind Society 5.0 and lists the relevant nomenclature. As per the Japanese government literature, Society 5.0 should be one that, "through the high degree of merging between cyber-space and physical space, will be able to balance economic advancement with the resolution of social problems by providing goods and services that granularly address manifold latent needs regardless of locale, age, sex, or language." The vision

The original version of this chapter was revised: This book was inadvertently published with the incorrect license type CC BY 4.0 and the Open Access License has been amended throughout the book to the correct license type CC-BY-NC-ND. The correction to this chapter is available at https://doi.org/10.1007/978-981-15-2989-4_9

A. Deguchi (✉)
Department of Socio-Cultural Environmental Studies, Graduate School of Frontier Sciences, The University of Tokyo, Tokyo, Japan
e-mail: deguchi@edu.k.u-tokyo.ac.jp

C. Hirai
Global Center for Social Innovation—Tokyo, Research & Development Group, Hitachi, Ltd., Tokyo, Japan
e-mail: chiaki.hirai.xj@hitachi.com

H. Matsuoka
Center for Exploratory Research, Research & Development Group, Hitachi, Ltd., Tokyo, Japan
e-mail: hideyuki.matsuoka.ws@hitachi.com

T. Nakano
Department of Housing and Urban Planning, Building Research Institute, Ibaraki, Japan
e-mail: nakano@kenken.go.jp

K. Oshima
Department of Urban Engineering, Graduate School of Engineering, The University of Tokyo, Tokyo, Japan
e-mail: oshimax@edu.k.u-tokyo.ac.jp

M. Tai
System Innovation Center, Research & Development Group, Hitachi, Ltd., Tokyo, Japan
e-mail: mitsuharu.tai.wu@hitachi.com

S. Tani
Social Systems Engineering Research Department, System Innovation Center, Research & Development Group, Hitachi, Ltd., Tokyo, Japan
e-mail: shigeyuki.tani.dn@hitachi.com

Society 5.0, https://doi.org/10.1007/978-981-15-2989-4_1

1

of Society 5.0 requires us to reframe two kinds of relationships: the relationship between technology and society and the technology-mediated relationship between individuals and society. With this perspective, the introductory chapter provides an overview of the concept of Society 5.0. It clarifies the differences between the society today and Society 5.0. It proposes how we approach Society 5.0 in this book.

Sections 1.1–1.4 of this chapter describe what is Society 5.0. In particular, the focus is on the following key concepts which are parallel aspects of the society: "a human-centered society," "merging cyberspace with physical space," "a knowledge-intensive society," and "a data-driven society." Understanding these four concepts enables us to develop the approach required to make Society 5.0 a reality. In Sect. 1.5, we clarify the conceptual differences between Society 5.0 and Germany's Industrie 4.0, which is one of the leading visions of revolutionizing the industry through IT integration. Society 5.0 seeks to revolutionize not only the industry through IT integration but also the living spaces and habits of the public.

Keywords Cyberspace and physical space · Data-driven society · Data literacy · Industrie 4.0 · Knowledge-intensive society

1.1 How We Approach Society 5.0

The Schema of Society 5.0

The basic schema of Society 5.0 is that data are collected from the "real world" and processed by computers, with the results being applied in the real world. This schema is not new in itself. To cite a familiar example, air-conditioning units automatically keep a room at the temperature programmed into the unit. An air conditioner regularly measures the room's temperature, and an internal microcomputer then compares the temperature reading with the registered temperature setting. Depending on the result, the airflow is activated or deactivated automatically, such that the room maintains the desired temperature. Many of the systems we rely on in society use this basic mechanism. It underlies the systems responsible for keeping our homes adequately supplied with electricity, and those that keep the trains running on time. This mechanism relies on computerized automated controls. When people use the term "information society," they mean a society in which each of these systems collects data, processes them, and then applies the results in a particular real-world environment.

So what makes Society 5.0 different? Instead of having each system operating within a limited scope, such as keeping a room comfortable, supplying energy, or ensuring that the trains run on time, Society 5.0 will have systems that operate throughout society in an integrated fashion. To ensure happiness and comfort, it is not enough just to have comfortable room temperatures. We require comfort in all aspects of life, including in energy, transport, medical care, shopping, education, work, and leisure. To this end, systems must gather varied and voluminous real-world data. This data must then be processed by sophisticated IT systems such as AI, as only these IT systems could handle such a vast array of data. The information yielded from such processing must then be applied in the real world so as to make our lives happier and more comfortable. But does this not happen already? The difference is that in Society 5.0, the resulting information will not just guide the operation of an air conditioner, generator, or railway; it will directly shape our actions and behavior. In summary, Society 5.0 will feature an iterative cycle in which data are gathered, analyzed, and then converted into meaningful information, which is then applied in the real world; moreover, this cycle operates at a society-wide level.

Merging Cyberspace and Physical Space

Having clarified the basic schema, we now turn to the next question: what do we mean by "merging the physical space (real world) and cyberspace?" Cyberspace refers to a digital space in which real-world data are collected and analyzed to derive solutions. The term was coined to describe an imaginary or virtual area, where swathes of raw data are freely accessed and converted into useful information, which can then be shared with others. The infrastructure of this space is the vast array of computer networks.

However, in the case of Society 5.0, cyberspace does not just mean a space for exchanging vast volumes of data. It also means a space created by computer networks for analyzing problems and modeling practical, real-world solutions. When the computer systems of Society 5.0 analyze raw real-world data, they must do so using a structure that mirrors the real, physical world. As complicated as this may sound, the principle is very simple. To use the air conditioner example again, the internal microcomputer runs a program to measure a variable that describes the room temperature (let us call this variable "T"). The program compares the T value against the registered temperature setting and then determines whether to activate or stop the airflow. Thus, such an air conditioner has a discrete cyber model that analyzes the room with a single parameter, T. Let us call this the "room model." Modern air-conditioning systems can also sense the positions of people in the room and customize the temperature accordingly. Such systems allow for a more complex cyber room model, one that uses a range of parameters—such as room size, temperatures of different parts of the room, and positions of the room's inhabitants. The more closely one wants to meet people's needs for happiness and comfort, the more granular (or closer to the real world) the cyber model must be (see Fig. 1.1). The ultimate objective of Society 5.0 is to incorporate real-world models into cyberspace such that they can deliver highly nuanced solutions to real-life problems.

What, then, is physical space? Physical space refers to the real world, from which raw data are collected and into which solutions are applied. Some might interpret "real world" to mean everything that is real, including computer systems. Hence, the government literature adopted the descriptor "physical" to distinguish this space from cyberspace. This book uses the expression "physical space (real world)."

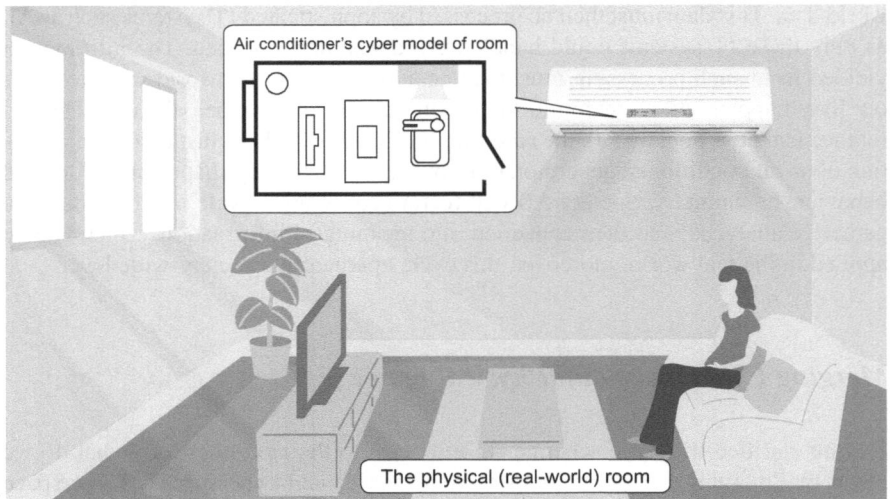

Fig. 1.1 Physical space (the room) and cyberspace (the air conditioner's model of the room)

As the next section explains, the idea of merging cyberspace with the physical space (real world) refers to a cycle in which data smoothly flow from the physical space (real world) into cyberspace and then flow back from cyberspace into the physical space (real world) in the form of meaningful information. Hitherto, we have relied on systems such as energy supply and rail transport systems, each of which governs some part of the physical world and is controlled separately. However, once all of these systems are interconnected through cyberspace, they will enable much more sophisticated services and produce much greater value in the real world.

Toward a People-Centric Society

It is through the mechanism described above that Society 5.0 will become a people-centric society. Originally, the purpose of an air conditioner was to keep a room at the desired temperature. The matter is simple enough if temperature control is our sole objective, but things start to get more complicated once our goal is a people-centric society. The government's 2017 comprehensive strategy describes a human-centered society as one that can "balance economic advancement with the resolution of social problems … to ensure that all citizens can lead high-quality lives full of comfort and vitality." The authors of the strategy described it as such because they understood how difficult it can be to balance economic development, resolution of social problems, and quality of life. Society 5.0 was thus proposed as a way to attempt this feat.

Air conditioners play an invaluable role in society; many offices and factories would struggle to function if their premises were not comfortably air-conditioned.

Yet air conditioners also contribute to global warming: they often run on power derived from burning fossil fuels, which releases greenhouse gases. Thus, we cannot only consider the need to keep buildings comfortably air-conditioned; we must also consider the effects upon society as a whole, or indeed upon our entire ecosystem. As this example illustrates, balancing these two interests is no easy task. If we single-mindedly pursue economic growth, we may end up becoming a society of mass production and mass consumption, and harm the planet in the process. However, if we forgo our pleasures and restrict our energy consumption to the bare minimum, life becomes drab and uncomfortable. Moreover, if we all lived such a spartan existence, the economy would stall. Society 5.0 is an attempt to overcome this seemingly intractable dilemma. In this book, we outline the approach to this dilemma, an approach that we have termed "Habitat Innovation." We also examine the direction of the technological developments underlying Habitat Innovation.

The task of solving social problems without sacrificing quality of life is difficult for another reason: it requires us to balance what is best for society with what is best for the individual. Suppose you live alone in a single-room apartment. Who decides on your air conditioner's temperature settings? Clearly, you are free to decide this for yourself. Suppose, however, that you are just one of the inhabitants. Each person may have their own temperature preferences. How do you ensure that you are all happy and comfortable? Should you take a poll of each person's preferred temperature and then calculate the mean? Should you hold a debate about the ideal temperature and then take a vote? Should someone in your group make a final decision? Not so simple anymore, is it? Yet this kind of scenario is at the easy end of the spectrum. Just imagine applying this to more complex social scenarios, in which you must consider the happiness of countless individuals, and do so using a dizzying array of scales and metrics. Could you reconcile or find an acceptable balance between the interests of the society and that of the individuals in it? This challenge is linked at a fundamental level to the question of what we mean by "high-quality lives full of comfort and vitality." There are many different definitions and measures of well-being. Well-being is not like the temperature of a room; you cannot quantify it in most cases. It will take us much more time until we can derive clear-cut solutions to this problem, but for the time being, humanities and social science researchers are delving into the peripheries of matter and considering how best we can approach the core.

The vision of society that Society 5.0 describes requires us to think about two kinds of relationships: the relationship between technology and society and the technology-mediated relationship between individuals and society.

1.2 Merging Cyberspace with Physical Space

In the previous section, we learned that the underlying mechanism of Society 5.0 is the merging of cyberspace with the physical space (real world). This section further clarifies what such a convergence means and how it can benefit society.

Modeling Real-World Issues

Cyberspace is the electronic world inside computers. Data from the physical space (real world) are analyzed in cyberspace so as to derive solutions for managing or improving society. Once these solutions are implemented in physical space (real world), the outcomes are evaluated, which generates data. This data is then input back into cyberspace for analysis and, if there are any problems, further solutions will be derived. This cycle, whereby society is continuously adjusted and improved, is what Society 5.0 is all about.

To derive solutions for the physical space (real world), cyberspace must have a structure mirroring that of the real world. Consider once again the example of the air conditioner (see Fig. 1.2). In this case, the cyber model must have a real-world mirroring structure necessary for air-conditioning the room. In other words, the system must model the physical characteristics of the room to understand how the room will change if the airflow is increased or decreased. If the system models the room's features as they are in reality, it can run cyber simulations and learn strategies for keeping the room optimally air-conditioned.

The impact of a given level of airflow upon the room temperature will depend on various factors, including the room's size, the heat-insulating properties of the walls, the number of inhabitants, and the exterior temperature. It is no easy task to acquire a model that accurately reflects the room's real-life conditions. This is where the Internet of Things (IoT) and artificial intelligence (AI) come in. IoT allows varied and voluminous data (in this case, the room's size, the temperatures in different parts of the room, the room's inhabitants and their spatial distribution, etc.) to be gathered

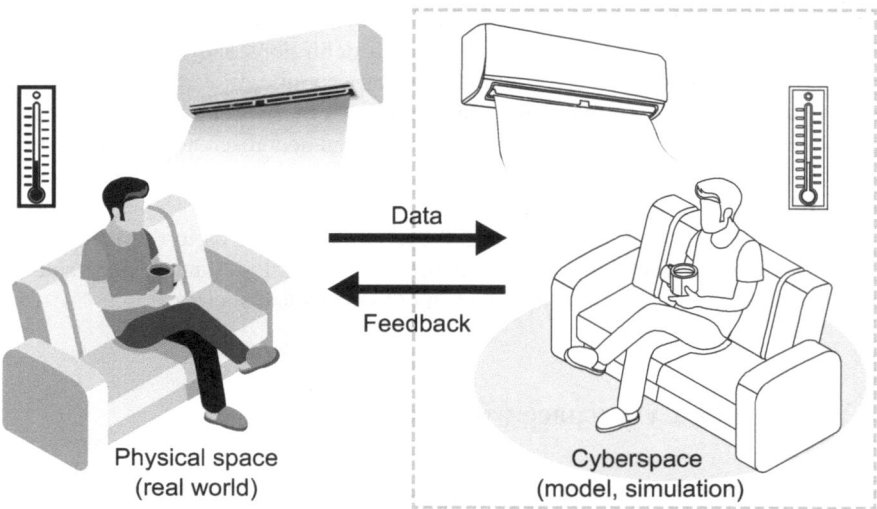

Fig. 1.2 Modeling the real world

in cyberspace. AI, on the other hand, can analyze the vast amounts of data obtained and then create a cyber model of the room that behaves just like the real thing.

Once this cyber model is established, the system can estimate how best to condition the room and then implement this strategy in the physical space (real world). The system can measure how the airflow is affecting the room temperature and incorporate this information back into cyberspace. If the room's actual temperature differs from the target temperature, then the cyber model of the room must have missed the mark. The AI notes the mistake and readjusts the model accordingly. Through this calibration cycle, the cyber model of the room will eventually come to adequately resemble the actual room. Thus, when the literature mentions the "merging" of cyberspace and physical space, it means that these two spaces have come to resemble one another so much as to be indistinguishable.

The idea of merging the cyber and the physical is not novel. Power generation and rail transport, for example, now use control systems that model their target environment so as to supply the right level of energy or run the trains on time. Such systems are known as cyber-physical systems (CPS). However, the convergence of the cyber and physical that Society 5.0 envisages does not involve separate, isolated systems. Society 5.0 is about cyber-physical convergence at the level of society as a whole. Convergence at this macro-level could perhaps be described as the merging of spaces with spaces.

Understanding How Services Are Interconnected

When the convergence comes to fruition, models that had until then been generated separately in each system will become interconnected in cyberspace. Consequently, we will come to see how different services interconnect. How will this insight benefit society?

We rely on many types of services, including those related to energy, transport, water, healthcare, public security, distribution, retail, education, and entertainment. It may appear that each service is separate, but they are in fact interconnected. To build a better society, we must learn to see how services interconnect and devise solutions accordingly.

Take urban traffic congestion as an example. One way of solving this problem might be to develop a subway system, but this costs time and money. Before rushing to take action, you should consider why the congestion occurs in the first place. In some cities, people prefer to travel by car because of poor public security. In other cities, the cause of congestion may be an inadequate water infrastructure, which causes roads to be inundated once it rains. In some cities, there is a rich riverine infrastructure, yet the inhabitants avoid the river bus owing to water pollution, a result of rapid urbanization. In other cases, congestion is the result of rampant illegal parking, which itself was caused by a failure to build adequate parking facilities close to marketplaces. As these examples illustrate, transport is interconnected with other services. Thus, although a subway system might be an effective solution to congestion, if the interconnections with other services are considered, a cheaper and quicker alternative, such as enhancing public security, installing better water infrastructure, improving sewage purification, or relocating marketplaces, may be discovered.

If an entire city is modeled in cyberspace, it will be possible to thoroughly analyze the root causes of the issue, which in this case is traffic congestion. It will also facilitate the process of devising solutions; simulations could be run in cyberspace to identify how best to allocate limited budgets so as to eliminate the congestion. The secondary effects of each potential solution could also be identified so as to avoid unintended consequences.

Urban planners already examine the relationship between different services. The difference is that the convergence of cyberspace and physical space (real world) will yield vast resources of data gathered from the physical space (real world). This data will help urban planners understand more accurately the interactions between different services. In other words, AI can spot connections that a human would overlook. With such AI, we will learn how different services in a given area interact in the short term, and how a given service would shape other services over a longer time span. Additionally, AI-derived insights into interservice dynamics may yield new services. In the years ahead, all these possibilities will garner more serious attention than they have received so far.

Thus, by coupling and linking in cyberspace services, which have been so far administered and managed separately, it will be possible to integrate services, and thus derive new value in the physical space (real world). This is the value we can expect to gain from connecting services via cyberspace.

Accumulating and Sharing Knowledge

Services are not the only things that can be linked in cyberspace. Cities can be linked with other cities, and societies with other societies. By modeling a city or society in cyberspace and linking it with other cities or societies, it will be possible to extrapolate existing knowledge.

Let us consider an example. Imagine that you have analyzed some data pertaining to a given city using a certain method. This method may be applicable to another city. As the two cities have different environments, the results of your analysis in the second city may have limited use in their raw form, but the analytical method itself is applicable to both cities. Now let us say that you implement a strategy in one city and record the outcome. Whether the strategy proves a success or a failure, the lessons could be applied to other cities in many cases. Likewise, case data on solutions to problems in Japan may be applicable to emerging nations, thereby crossing physical and temporal barriers.

As mentioned previously, "cyberspace" originally meant an imaginary or virtual space wherein vast sums of raw data are freely and broadly accessed and converted into meaningful information, which then gets shared among or viewed by different users. As also mentioned previously, the infrastructure upon which cyberspace

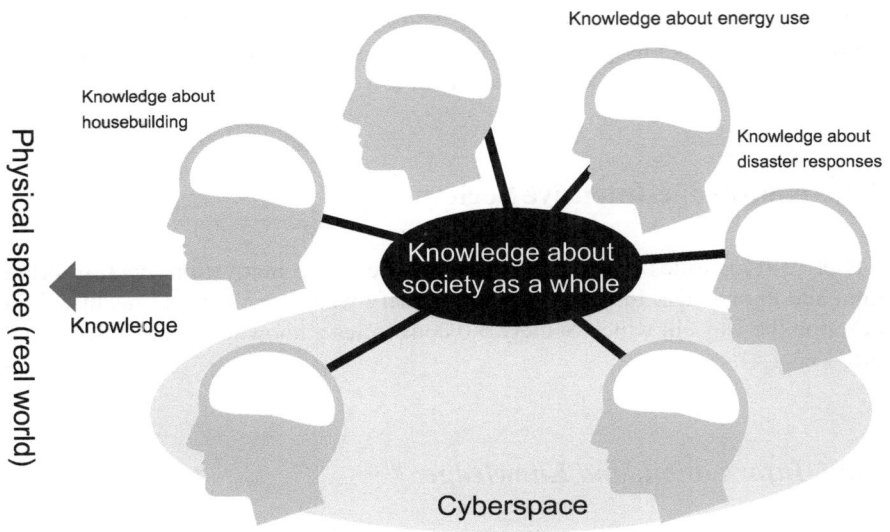

Fig. 1.3 Accumulating and sharing knowledge

exists is the vast array of computer networks. These computer networks enable information and knowledge to be shared without the restrictions of time and space. This accumulation and sharing of knowledge is the original purpose of cyberspace (see Fig. 1.3).

There are many ways in which cyberspace could facilitate the accumulation and sharing of knowledge, in addition to modeling and analyzing phenomena in physical space (real world). For example, if a municipality succeeds in becoming a supersmart society, the knowledge behind this success can be applied the very next day in another municipality situated far away. Then, some decades later, the knowledge can be used overseas, in a country that is less economically developed.

In this section, we discussed what the "merging of cyberspace and physical space (real world)" means in the context of Society 5.0. We also discussed how cyberspace can help us link together real-world phenomena so as to create new value. The "merging" refers to the process of gathering raw data from the physical space (real world), using the data to derive models in cyberspace, and iteratively improving these models. This process creates value in that the models generate new knowledge, which can then be accumulated and shared. It differs from the existing process in that a much broader array of data is gathered, and gathered at a much greater volume and a much higher frequency by comparison. Another difference is that AI and other modern innovations can process the vast ocean of data to derive new knowledge.

Insofar as we focus on the knowledge-production aspect, we might aptly call Society 5.0 a knowledge-intensive society. If we focus more on the data-production aspect, we might want to call it a data-driven society. So far, we have not clearly defined the terms "data," "information," and "knowledge." The following section, however, clarifies our usage of each of these terms and then discusses what we mean by a knowledge-intensive society and a data-driven society.

1.3   Knowledge-Intensive Society

Society 5.0 identifies three elements that drive social innovation: data, information, and knowledge. In this section, we clarify what these three terms mean and describe the ways in which Society 5.0 constitutes a knowledge-intensive society (see Fig. 1.4).

Data, Information, and Knowledge

First, what are data? Generally, data refer to tangible and intangible phenomena in the physical space (real world) that are represented as numerical values, states, names, or binary figures (0 or 1) telling us whether a thing is present or absent. To illustrate this definition, we will refer to the population of a hypothetical municipality (let us call it Town A). In Japan, the town's population could be worked out by referring to the relevant entries in the national registry of citizens (the "Basic Resident Register"). From this source, the attributes of Town A's residents, including their gender, household composition, and address, could be found. These facts rep-

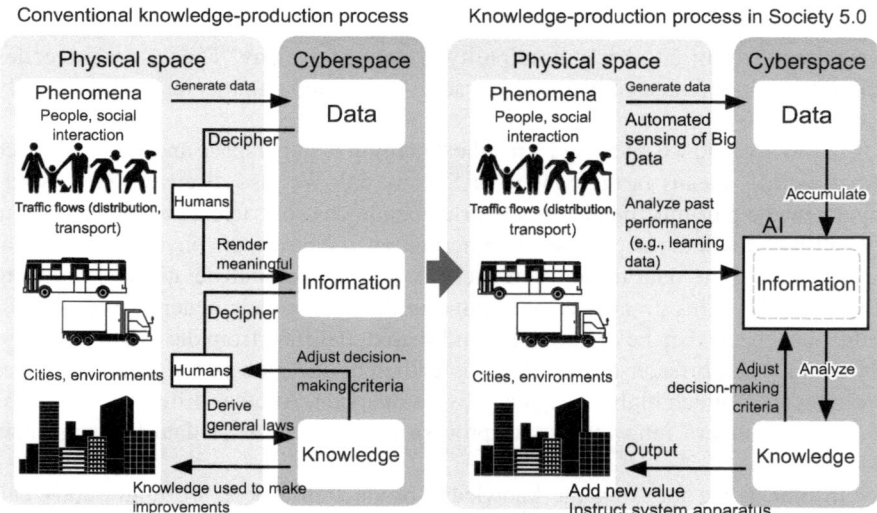

Fig. 1.4 Data, information, and knowledge

resent Town A's *data*. Data are the most basic of the three elements (data, information, and knowledge) that are accumulated in cyberspace.

If this is data, what is information? Information is data that has been rendered meaningful by selecting and processing it for a particular purpose or as part of a course of action. To return to Town A, once you have the raw population data, it could be broken down by age group to see the demographic trends over the past 10 years or the rate of aging. The age breakdown could also be used to plot a graph showing the population pyramid. The results of such analysis represent Town A's *information*. By analyzing the demographic trends, you could determine whether Town A is on a growth trajectory (its population is growing) or whether it is on the decline (its population is shrinking). It is the addition of such meaningful indications that turns data into information.

Suppose the information tells you that Town A's population is shrinking. To address this problem, you must analyze the causes of the population decline. Perhaps the decline is driven by falling birthrates and population aging. Or perhaps there is a net outflow (the people moving away from the town outnumber the people coming in). The decision of what to do could be worked out by comparing Town A's population trends with that of other municipalities and referring to best practice models developed by experts. *Knowledge*, then, is what enables you to make a decision. Information becomes knowledge when it is comprehended, analyzed, and related to general laws, including best practices and precedents. Knowledge can also be described as generalized observations extracted from individual cases. Knowledge allows you to surmise the causes of a problem, and it also helps you to derive solutions to address these causal factors. The more knowledge you have, the more equipped you are to derive a judicious information-based decision.

What Is a Knowledge-Intensive Society?

Data becomes useful to us once we convert it to information, and then into knowledge. Hitherto, this conversion process has been driven by human–computer interactions. In Society 5.0, the process will be driven without human intervention; of the three elements, humans will only gain greater opportunities to access AI-derived knowledge, the final output of the conversion process.

How will this change affect society?

Like other developed nations, Japan evolved from a labor-intensive society, in which production relied on the efforts of a massive workforce, into a capital-intensive society, which was focused on tangible goods and was based on mass production and mass consumption (both of which resulted from industrial revolutions). In the capital-intensive society, cities developed around seaports and airports where tangible goods were clustered. Under the Society 5.0 way of thinking, however, value is generated not from clusters of tangible assets but rather from knowledge spaces—spaces where data and information are gathered and then deciphered and deployed through knowledge (Gonokami 2017). In this sense, a knowledge-intensive society is a key aspect of Society 5.0.

New knowledge will arise when data and information are deployed interconnectedly. New knowledge can spark innovation in tertiary industries such as services, but it will also do so in the more traditional primary and secondary industries such as agriculture and manufacturing. Japan's agricultural sector is somewhat inefficient owing to sporadically distributed farmland. A knowledge-intensive Japan, however, could spark an agricultural renaissance by leveraging detailed spatial information and predictive weather knowledge along with drone and robotic technologies. A knowledge-intensive society may also generate new industries and transform the industrial structure.

In pursuing this paradigm shift, universities and businesses, which have until now played a core role in technological development, will need to play a new kind of role. The role of technology thus far has been to add value to tangible goods, but in the knowledge-intensive society, universities and businesses will need to help cultivate new industries, which in turn will generate new value by clustering and combining knowledge.

Rules and Norms in the Knowledge-Intensive Society

In the coming knowledge-intensive society, technology will play a critical role in building information integration architecture—architecture that enables data to be collected, synthesized, and then integrated with information in heterogeneous fields. At the same time, however, we must establish rules and norms governing how we approach data. Data producers must uphold certain rules and standards of conduct, and those who analyze or use the data must be sufficiently data literate.

Let us consider the situation for data producers. Technology facilitates the knowledge-generation process, but no matter how advanced this process becomes, if the data is unsuitable for analysis, you will fail to derive accurate knowledge. Although automated processes can catch some data errors, it is difficult at present to catch every error owing to the lack of a coordinated system. In other words, every data producer follows its own separate method of data production. To illustrate this point, we will use a familiar example: tourism. Until 2009, when the Japan Tourism Agency issued the Common Standards for Estimating Tourist Arrivals (Japan Tourism Agency 2019), each municipality followed its own method to survey and compile tourist data. This practice prevented the data from being useful; although tourist trends could be analyzed in each municipality, the trends between municipalities could not be compared. Another issue was that despite the incomparability of the data, third parties might attempt comparisons anyway, which would result in erroneous knowledge. If anyone can tally the number of visitors with a simple device and then publish the data online, it is all the more important to establish common standards and procedures, so that data producers approach the data judiciously, understanding how it will be used.

Information Literacy

What about the people who analyze and use the data? One of the top tasks in relation to Society 5.0 is to ensure that such individuals are literate in personal data and information. Let us consider an example. As part of the European Union's Horizon 2020 program (European Commission 2019), Barcelona organized the "smart citizen" project, in which citizens developed a sensor board that can be installed in balconies to monitor air and noise pollution. The data recorded by the sensors is published as open-source data (Smart Citizens 2019), and citizens can cite this open-source data in their campaigns for better environmental policies. In this project, Barcelonans are the data producers, and insofar as they derive meaningful information from the data, they are data users as well. By contrast, Japanese people typically regard data use as the sole preserve of public servants and businesses, and few see data as something that they themselves could use, as Barcelona's "smart citizens" do. What matters is to promote public discussion and action regarding the society-wide use of data.

The benefits of Society 5.0 should be enjoyed by all. As the Japanese government literature says, Society 5.0 should be one that, "through the high degree of merging between cyberspace and physical space, will be able to balance economic advancement with the resolution of social problems by providing goods and services that granularly address manifold latent needs regardless of locale, age, sex, or language." But can you have too much of a good thing? If every service and business is highly data driven, might this not encourage people to lose their agency in society and passively follow AI-generated recommendations on which goods to purchase or which services to use? That does not sound like a very interesting life. If the goods and services of society are to be available to all, we must ensure that people still lead purpose-driven and creative lives. To this end, universities and businesses will have an increasingly crucial role to play. As we move toward a truly people-centric life, progress in information technology must be accompanied by efforts to train up industrial innovators and raise the information literacy of each and every citizen. Universities, for their part, in addition to spurring technological progress as before, must additionally be responsible for cultivating literacy among information users through both general curricula and recurrent education, so as to promote the civil society that embodies Society 5.0.

1.4   Data-Driven Society

Society 5.0 is described as a data-driven society. What is a data-driven society? We live in a so-called information society, so how does this differ from a data-driven society? The previous section defined information as data that has been processed and rendered meaningful, while it defined knowledge as the general empirical laws extracted from such information. Compared to information and knowledge, data exist at a more basic level. What, then, does it mean for a society to be driven by this most primitive of the three elements? The data-driven society is crucial for under-standing Society 5.0, a point that is aptly illustrated by the fact that both terms appear in the Japanese Government's "Growth Strategy 2018" (Growth Strategy Council 2018). Accordingly, this section explores the question in some detail.

What Is a Data-Driven Society?

First, let us see how the data-driven society is defined in the government literature. The term featured in the literature even before Society 5.0 was proposed. For exam-ple, it appeared in a 2015 report of the Ministry of Economy, Trade and Industry's (METI) Industrial Structure Council (Ministry of Economy 2015). This report defines a data-driven society as a society "where the above-mentioned CPS is applied to various industrial societies through digitization and networking of things using IoT, and the digitized data is converted into intelligence and applied to the real world, and then the data acquire added value and move the real world [sic]." In this quotation, "intelligence" equates with the information and knowledge discussed in the previous section.

More simply, the data-driven society is a society where data (gathered by IoT networks) are converted into information and knowledge, which then "drive" (or as the literature says, "move") the real world. As accurate as this definition may be, it may still leave readers nonplussed. The previous section described the relationships between data, information, and knowledge, but this does not give us a clear picture of how data drives the real world. So how exactly does data drive the real world? It drives the world in two different ways. First, data drives the world indirectly via humans. That is, vast resources of data inform and guide human decision-making, which then effects change in the world. Second, data drive the world directly (with-out the mediation of humans) through automated processes. Let us consider exam-ples of both.

Regarding the former, suppose you are designing an urban transport system; under a conventional approach, you would consult data and then make decisions based on this data. You would rely on numerous researchers to gather traffic volume data using manually operated head counters, and these findings would inform your designs for road traffic, bus services, metro system, and the like. However, because these traffic data are costly to gather, only a limited amount are available (there are only data for a limited number of sites in the city and these are dated several years apart).

In the data-driven society, however, the data available would be staggering in volume and breadth, and be real-time data to boot. Technology allows you to moni-tor the traffic flows across the city as a whole in real time. For example, to monitor people flows, you could refer to smartphone data or access the data of prepaid trans-port cards (known as IC cards in Japan). To monitor foot and vehicle traffic volume, you could analyze the footage of CCTV cameras installed along roads and in buildings. You could also collate this data with shopping data to gain insights into the motives for people's movements. By visually modeling all this urban data in real time, you will grasp the entire workings and dynamics of the city.

Before enacting any changes in the city, you must hold a consultation process in which numerous stakeholders share their understanding of the status quo and how it should be changed, if at all. A visual model of the city grounded in voluminous, varied, and real-time data would radically shape this consultation and decision-making processes. This is what it means for data to drive society indirectly, via humans.

Now for the latter meaning—a society that is driven directly by automated systems. One example of automated control systems is traffic signals. Traffic lights shift between red, amber, and green, thanks to the operation of an internal computer program, one that humans designed.

However, if we want the kind of people-centric society that Society 5.0 describes, we must consider numerous variables and needs, even if we limit our focus to a traffic control system. Drivers may want minimal congestion, residents may want minimal traffic flows so as to limit exhaust fumes, and pedestrians might wish to have minimal waiting times at crosswalks. Railway level crossings can be a source of traffic congestion, so rail timetables would also have to be considered. All in all, a traffic control system is a very complex matter.

It is all but impossible for humans to design a program that can control traffic signals absolutely optimally, taking into account all the above variables and needs. Hence, we must look to AI. Humans can define an optimal traffic state and then let AI coordinate traffic signals accordingly. If we regularly input data, such as traffic volumes, exhaust volumes, and pedestrian waiting times, AI will start to learn the outcomes it can expect from a given traffic control pattern. In this way, AI will progressively derive general laws on how best to control traffic. Over time, the AI will learn how transport is affected by factors such as public events and weather conditions and come to understand the optimum responses to such phenomena.

Thus, in the future, AI will convert data into knowledge (general empirical laws) through an automated process, and then use this knowledge to automatically control traffic. Instead of traffic signals being controlled by a human-made computer program, they will be controlled by AI-generated optimum algorithms. This process is mediated by data, but not by humans: that is the second meaning of a data-driven society.

From the Information Society to the Data-Driven Society

So far, we have learned that the data-driven society is a society where IoT-gathered data is converted into information and knowledge, which then drives the real world either indirectly (with the mediation of humans) or directly (through automation). How does this differ from an information society? An information society derives value from information. A data-driven society (in both senses) derives value from data. The government's Growth Strategy 2018 (Growth Strategy Council 2018) describes this idea in stark terms:

> "…in the data-driven society of the 21st century, the most important currency of economic activity is high quality, up-to-date and abundant 'real data'. Data has become so valuable that saying that the success or failure of a business depends on its access to data [is] by no means an exaggeration."

Some might argue that we should shorten the term "data-driven society" to "data society," so as to more easily compare and contrast it with the "information society." However, the government decided to add "-driven" to underscore how future technological progress will result in extensive automation (nonhuman-mediated processes).

In this section, we learned about the two ways in which society will be data driven. Of the two, an automated society may seem the more futuristic. However, it would be a mistake to think of a human-mediated society as a transitionary state between today's society and the ultimate state of full automation. Instead, human mediation and automation will exist side by side. In the case of traffic signals, AI is responsible for effectuating an optimal state, but it is humans who decide what this state is in the first place. Human-mediated processes, such as consultations in which the participants refer to visual urban data, will play an ever-greater role in building the people-centric society. We are the ones who decide how to strike a balance between different comfort needs, such as between drivers' desire to travel smoothly without needing to constantly stop at red lights and pedestrians' desire to cross the road quickly. Likewise, it is humans who define the criteria for measuring comfort and happiness. Standards of happiness vary between cultures and time periods. To find the right balance, consultation processes should involve as many stakeholders as possible, not least of whom should be residents—the chief actors of a local community. Once full consultations have been made and a consensus reached, this consensus can then be put into effect by automated technology. These parallel aspects of a data-driven society, by operating in tandem in this way, will support the people-centric Society 5.0 and provide the flexibility necessary to ensure that the underlying architecture is applicable in many different countries and cultures. Thus, solutions generated in Society 5.0 can contribute to other social problems in different parts of the world.

1.5   Industrie 4.0 and Society 5.0

In November 2011, the German Federal Government released "High-Tech Strategy 2020 Action Plan for Germany" (Industrie 4.0 Working Group 2013), which outlined a high-tech strategic initiative called Industrie (Industry) 4.0. This vision predated Society 5.0, as proposed in the 2016 Science and Technology Basic Plan, by 5 years. Why did Germany pursue a national campaign to promote science and technology in its manufacturing sector? This section outlines the new industrial vision that Industrie 4.0 encapsulated. It also compares Industrie 4.0 with Society 5.0 as a means of further clarifying the latter.

What Was Industrie 4.0?

Industrie 4.0 was a national strategic initiative led by the Ministry of Education and Research (BMBF) and the Ministry for Economic Affairs and Energy (BMWI). To deliberate on the initiative, a working group was formed consisting of actors from government as well as from businesses and universities. The working group was led by Henning Kagermann, former chairman of SAP SE and president of the German Academy of Science and Engineering (acatech). In April 2013, the working group issued its recommendations in a report titled "Recommendations for implementing the strategic initiative INDUSTRIE 4.0" (Industrie 4.0 Working Group 2013).

The report focused on deploying IoT in manufacturing so as to enable cyber-physical (CPS) systems that can add value to production activities. It also focused on promoting "smart factories," which are factories that achieve significant savings in manufacturing costs.

According to the report, smart factories should use IoT devices and the Internet to gather data on all stages of the production process in the physical space (real world), and then recreate this data in cyberspace. AI then analyzes this cyber data, or runs simulations to derive optimal solutions. AI's findings will be automatically fed back into real-world factory control systems. Simply put, smart factories are factories that think for themselves.

Smart factories enable automation and optimization across all aspects of manufacturing. As well as managing general production processes, they could handle payments for parts; they could even detect any abnormalities or deficiencies in the production apparatus and then automatically fix the problem or recalibrate a process. The chief actors in smart factories are sensors and AI.

As a proper noun, Industrie 4.0 denotes a uniquely German initiative, but the underlying concept—to deploy IoT in manufacturing—has gained global traction. This concept is more generally described as the "fourth industrial revolution," and it describes an extensive trend to overturn industrial production.

But why *four*? To understand this, we need to recap the history of industrialization (see Fig. 1.5).

The first industrial revolution began in Britain in the eighteenth century, and it was driven by the mechanization of manufacturing equipment. Water- and steam-powered machinery enabled a leap in productivity in the textile industry and other industries. The second industrial revolution began around the turn of the twentieth century, and involved mass production based on the division of labor. Producers shifted to fossil fuel-generated electric power, and factories became much larger. This second industrial revolution was epitomized by the Ford Motor Company's auto production. The third industrial revolution, which began during the 1970s, involved electronics. Producers used robotic technology to automate some manufacturing processes, and consequently achieved significant leaps in productivity. It was during this time that Japanese manufacturing gained worldwide prominence.

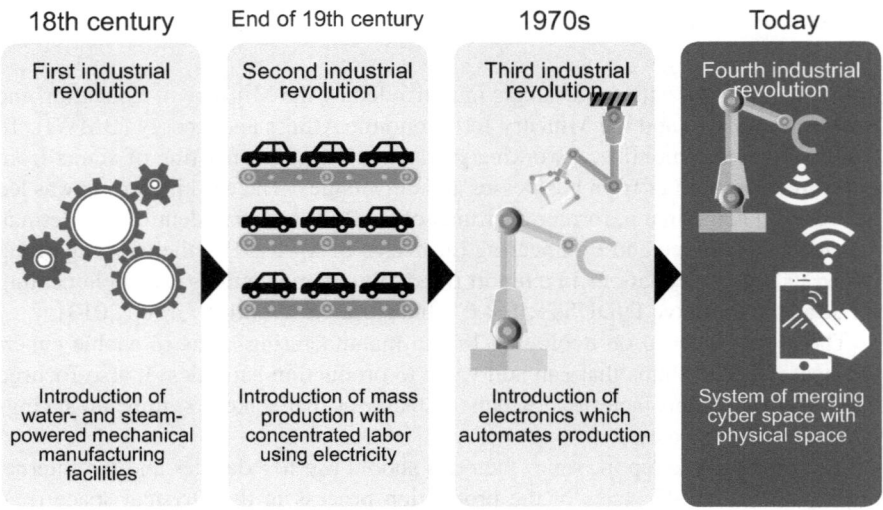

18th century	End of 19th century	1970s	Today

Fig. 1.5 The chronology of the industrial revolutions and the position of the fourth industrial revolution

Industrie 4.0 heralds the next stage of industrialization. As many readers will know, Japanese manufacturers already use robotics and sensor technology, and many processes are automated. Many of these readers may feel that the Japanese manufacturing has already made great strides in terms of productivity. Yet Industrie 4.0 is not just about making factories more efficient. As Taro Yamada argues, Industrie 4.0 is all about creating a data–information–knowledge cycle, in which all manner of manufacturing-related data, including data related to designs, clients, and suppliers, are gathered and shared among different fields and organizations (Yamada 2016).

The key difference between the third and fourth industrial revolution is that the latter uses data in a manner that surpasses traditional manufacturing frameworks. In the past, data related to the use of products, for example, would be abandoned upon the sale of the products; in the fourth stage of industrialization, however, manufacturers continue to gather this data after the products are sold. This practice allows manufacturers to identify latent needs from clients' Big Data and strengthen their value networks, thereby creating new business opportunities. Another difference with Industrie 4.0 is that added value is created through mass customization. In other words, AI drives customized output, flexibly accommodating diverse demand.

Although Industrie 4.0 focused primarily on manufacturing, the scope of the project extends farther. The vision requires the establishment of data-related standards and regulations (as well as the institutional environment necessary for such), which necessitates a collaborative process involving not only core manufacturing industries, such as the auto and electronics industry, but also IT and communications industries, academia, and government. Industrie 4.0 was not the first project to propose information integration. In 1984, Ken Sakamura of the University of Tokyo launched an open architecture real-time operating system kernel design called

TRON (*The Real-time Operating System Nucleus*) Project. In the 1987 and 1988 proceedings of the TRON Project, the concept of a "highly functional distributed system" (HFDS) was proposed (Sakamura 1988). Likewise, the phrase "Internet of Things" predates Industrie 4.0. Kevin Ashton, founder of the Auto-ID Center at the Massachusetts Institute of Technology, writes, "I'm fairly sure the phrase 'Internet of Things' started life as the title of a presentation I made at Procter & Gamble (P&G) in 1999." Ashton also clarifies that he uses the term to underscore the importance of linking intangible information with physical "things" (Ashton 2009). Thus, the idea of information integration architecture predated Industrie 4.0's launch in 2011, and businesses and academics were already pursuing their own research projects in this area. The role played by the Industrie 4.0 initiative was to reaffirm the importance of such innovation. Industrie 4.0 was proposed as a top-down national strategy involving collaboration between industry, academia, and government. Such an approach was necessary because the task of building an information integration architecture among industry, academia, and government represented the core of the "fourth industrial revolution," one that holds the key to innovating in manufacturing and industry, in general. Japan has taken a similar approach. In March 2017, Hiroshige Seko, Minister of Economy, Trade and Industry, attended the German computer expo CeBIT in Hannover and declared the government's vision of "connected industries" (Ministry of Economy 2017).

What Are the Aims of Industrie 4.0 and Society 5.0?

The aims of Industrie 4.0 were outlined in the German Federal Government's High-Tech Strategy 2020 Action Plan for Germany, the German equivalent of Japan's Science and Technology Basic Plan. So how is Industrie 4.0, as outlined in High-Tech Strategy 2020 Action Plan for Germany, compared with Society 5.0, as outlined in the fifth Science and Technology Basic Plan? As Fig. 1.6 illustrates, there are some commonalities. Both visions emphasize the use of technology, including IoT-related technology, AI, and Big Data analysis. Similarly, they both entail a top-down, state-led approach with collaboration between industry, academia, and the governmental sector.

There are some differences, however. Industrie 4.0 advocates smart factories, while Society 5.0 calls for a supersmart society. In addition, although both visions advocate the deployment of cyber-physical systems, the scope of deployment differs; in Industrie 4.0, CPS is to be deployed in the manufacturing environment, while in Society 5.0, it is to be deployed across society as a whole.

The two visions also differ in terms of measuring outcomes. Industrie 4.0 aspires to create new value and minimize manufacturing costs. Such down-to-earth outcomes allow for relatively simple and clear-cut performance metrics. By contrast, Society 5.0 aspires to create a supersmart society. The metrics in this case are much more complex. According to the Comprehensive Strategy on Science, Technology and Innovation for 2017, success is to be measured by how far society

Title	Industrie 4.0 (Germany)	Society 5.0 (Japan)
Design	•High-Tech Strategy 2020 Action Plan for Germany (BMBF, 2011) •Recommendations for implementing the strategic initiative INDUSTRIE 4.0 (Industrie 4.0 Working Group, 2013)	•5th Science and Technology Basic Plan (released 2016) •Comprehensive Strategy on Science, Technology and Innovation for 2017 (released 2017)
Objectives, scope	• Smart factories • Focuses on manufacturing	• Super-smart society • Society as a whole
Key phrases	•Cyber-physical systems (CPS) •Internet of Things (IoT) •Mass customization	• High-level convergence of cyberspace and physical space • Balancing economic development with resolution of social issues • Human-centered society

Fig. 1.6 Industrie 4.0 vs. Society 5.0. Source: Produced by authors

can "balance economic advancement with the resolution of social problems by providing goods and services that granularly address manifold latent needs regardless of locale, age, sex, or language to ensure that all citizens can lead high-quality, lives full of comfort and vitality" (Cabinet Office 2017).

There is also considerable difference in the scope of the intended future effects of technological innovations. Industrie 4.0 calls for an industrial revolution centered on manufacturing, but says nothing about how such a revolution may impact the public. By contrast, as illustrated by its concept of a people-centric society, Society 5.0 focuses heavily on the public impact of technology and on the need to create a better society. Included within the scope of Society 5.0's vision is a course of reform intended to engender an inclusive society that caters to diverse needs and preferences. This important differentiating aspect of Society 5.0 was mentioned in an address delivered by Prime Minister Shinzo Abe to Chancellor Angela Merkel during the CeBIT conference in Hannover. Upon hearing Abe's statements about Society 5.0, Merkel expressed her strong support for the vision (Prime Minister's Office of Japan 2017; JETRO 2017a, b).

The Common Issues for Both Industrie 4.0 and Society 5.0

Japan is sometimes said to be a problem-stricken first-world country. The problems that Japan faces are complexly interwoven such that an improvement in one area often comes at the cost of another. To give an example, curbing welfare spending might be good for the nation's fiscal health, but it would lead to grave problems in medical and healthcare environments. Similarly, we all understand the need to cut carbon emissions, but if we must live frugal lives to minimize their carbon footprint, that would run counter to the goal of ensuring that "all citizens can lead high-quality lives full of comfort and vitality."

Accordingly, to ensure that Society 5.0 can solve these dilemmas and create a people-centric society, it is necessary to clarify the target metrics of such a society as well as the roles that policy and technology should play in achieving them. Chapter 2 of this book goes into more detail on the metrics for different social issues, including those related to a carbon-free society and the health of the elderly.

Industrie 4.0, with its vision of smart factories, emphasized the manufacturing sector as the main physical space (real world); as for cyberspace, it envisaged a CPS-centered cyber architecture wherein information is integrated horizontally between different industries and vertically within manufacturing systems. On the other hand, Society 5.0, with its vision of a supersmart society, emphasizes society as the main physical space (real world); as for cyberspace, it must strive for a CPS-centered cyber architecture wherein information is integrated horizontally between different service sectors (e.g., energy, transport) and vertically within the systems that track each service user's history and attributes (such as their medical information, consumption behavior, and educational history). It must also achieve solid information security to enable the use of information.

Both Society 5.0 and Industrie 4.0 reflect Japan and Germany's responses to global initiatives, and both make a statement to the international community. Both visions seek the integration of information between different industries or sectors, and they both face the same challenges to such an end: the need to overcome the regulatory and technical bottlenecks that stand in the way of constructing the necessary cyber architecture, and the need to establish ISO-style international standards and international information security institutions, which are necessary for building such an architecture. Many commentators note that Western countries lead the way on this score, so Japan must press ahead with building an information integration architecture, while keeping an eye on global trends. Both Industrie 4.0 and Society 5.0 seek to build global cyber architecture that can serve as a safe environment for creative activities. A key factor that will determine their success in achieving this goal will be how well they work with Western countries, China, and the international community at large.

In the case of Society 5.0, one key challenge concerns how to optimally balance the needs of society with the needs of the individual. We cannot achieve progress until we solve this problem. The actors involved in policy and technology must coordinate with each other so that everyone understands how each policy proposal or technological development fits into and contributes toward Society 5.0. Otherwise, these actors will pursue their own particular technologies or policies in an uncoordinated fashion without understanding how they fit into the larger picture of Society 5.0.

In relation to this challenge, Chap. 2 clarifies the main social issues that Japan faces and outlines a framework for addressing them—namely, Habitat Innovation. Whereas Germany's Industrie 4.0 focused on industry, Society 5.0 envisages a future society. In other words, in addition to revolutionizing industry through IT integration, Society 5.0 seeks to revolutionize the public's living spaces, or habits. Further progress must be made in promoting applied smart city initiatives. Additionally, the policies necessary for optimizing society (so as to solve social issues) must be adeptly linked with the technology necessary to deliver high-quality social services (that enable the public to live happy, comfortable lives). With this in mind, we have presented tentative suggestions for balancing the interests of society with those of individuals.

References

Ashton K (2009) That "Internet of Things" thing: in the real world, things matter more than ideas (article on RFID Journal website), July 22, 2009. https://www.rfidjournal.com/articles/view?4986. Accessed 4 June 2019

Cabinet Office (Council for Science, Technology and Innovation) (2017) Comprehensive strategy on science, technology and innovation (STI) for 2017 (released on June 2, 2017), p 2. https://www8.cao.go.jp/cstp/english/doc/2017stistrategy_main.pdf. Accessed 4 June 2019

European Commission (2019) Horizon 2020. https://ec.europa.eu/programmes/horizon2020/en/. Accessed 4 June 2019

Gonokami M (2017) Society 5.0 (chishiki shūyakugata shakai) e no shakai henkaku to daigaku no yakuwari (Social innovation aimed at Society 0.5 <the knowledge-intensive society> and the role of universities), reference material used by the Ministry of Finance (Fiscal System Subcommittee, Fiscal System Council), October 2017. https://www.mof.go.jp/about_mof/councils/fiscal_system_council/sub-of_fiscal_system/proceedings/material/zaiseia291004.html. Accessed 4 June 2019

Growth Strategy Council (Headquarters for Japan's Economic Revitalization) (2018) Growth strategy 2018, June 2018. https://www.kantei.go.jp/jp/singi/keizaisaisei/pdf/miraitousi2018_en.pdf. Accessed 4 June 2019

Industrie 4.0 Working Group (2013) Recommendations for implementing the strategic initiative INDUSTRIE 4.0: final report of the Industrie 4.0 Working Group, April 2013. https://www.din.de/blob/76902/e8cac883f42bf28536e7e8165993f1fd/recommendations-for-implementing-industry-4-0-data.pdf. Accessed 4 June 2019

Japan Tourism Agency (2019) Kyōtsū kijyun ni yoru kankō irikomi kyaku tōkei (Tourist estimates based on common standards). http://www.mlit.go.jp/kankocho/siryou/toukei/irikomi.html. Accessed 4 June 2019

JETRO (2017a) Participation in CeBIT 2017 with largest pavilion ever (article on JETRO website), March 2017. https://www.jetro.go.jp/en/jetro/topics/2017/1703_topics3.html. Accessed 4 June 2019

JETRO (2017b) Abe Shushō, shakai o sumātokasuru "sosaetī 5.0" o teishō: IT Mihonichi "CeBIT 2017" ni pātonaākantorī to shite sanka (Prime Minister Shinzo Abe proposes Society 5.0 as model for making society smarter: Japan attends IT trade fair CeBIT 2017 as partner country) (article on JETRO website), April 2017. https://www.jetro.go.jp/biznews/2017/04/2e50a128af33afd2.html. Accessed 4 June 2019

Ministry of Economy, Trade and Industry (2017) "Connected Industries" as a goal that Japanese industries should aim for, March 2017. http://www.meti.go.jp/english/press/2017/0320_001.html. Accessed 4 June 2019

Ministry of Economy, Trade and Industry (Information Economy Subcommittee, Distribution and Information Committee, Industrial Structure Council Commerce) (2015) Interim report: changes in response to the arrival of a data-driven society using CPS, May 2015. http://www.meti.go.jp/shingikai/sankoshin/shomu_ryutsu/joho_keizai/pdf/report01_04_00.pdf. Accessed 4 June 2019

Prime Minister's Office of Japan (2017) Address by prime minister Shinzo Abe at CeBIT Welcome Night March 19, 2017 (article on government website). https://japan.kantei.go.jp/97_abe/statement/201703/1221682_11573.html. Accessed 4 June 2019

Sakamura K (ed) (1988) TRON purojekuto '87–'88 (TRON Project 1987, 1988), Personal Media Corp., pp 3–19

Smart Citizens (2019). https://smartcitizen.me/. Accessed 4 June 2019

Yamada T (2016) Nihongoban indasutorī 4.0 no kyōkasho: IoT jidai no monozukuri senryaku (A Japanese-language textbook on Industrie 4.0: a manufacturing strategy for the IoT era). Nikkei Business Publications

Chapter 2
Habitat Innovation

Hideyuki Matsuoka and Chiaki Hirai

Abstract Society 5.0 balances the best interests of the society as a whole, which involves the resolution of social issues, with the best interests of individuals, which is the indication of a human-centered society. In this chapter, we discuss the key performance indicators (KPI) formula as an approach to balancing these two factors. Under the context of "Habitat Innovation," the following approach is proposed to address social issues. In Habitat Innovation, the KPIs are factorized into three components that are "structural transformation," "technological innovation," and "quality of life (QoL)." Government leadership is required for "structural transformation." This component suggests ways in which the cyber-physical convergence framework can be deployed in the policymaking process. The "technological innovation" component tells us how the cyber-physical convergence framework can help to create a resource-efficient society. The "QoL" component can prompt us to deploy data in a way that generates new services for supporting people's QoL. In Habitat Innovation, the insights of engineering, social sciences, humanities, and many other disciplines are used to analyze what QoL means at an individual level and to identify the role that policy and technology should play in enhancing it. Examples using the Habitat Innovation framework to solve key social issues are shown.

Keywords Key performance indicator (KPI) · Quality of life (QoL) · Social issue · CO_2 emission · Technological innovation

The original version of this chapter was revised: This book was inadvertently published with the incorrect license type CC BY 4.0 and the Open Access License has been amended throughout the book to the correct license type CC-BY-NC-ND. The correction to this chapter is available at https://doi.org/10.1007/978-981-15-2989-4_9

H. Matsuoka (✉)
Center for Exploratory Research, Research & Development Group, Hitachi, Ltd., Tokyo, Japan
e-mail: hideyuki.matsuoka.ws@hitachi.com

C. Hirai
Global Center for Social Innovation—Tokyo, Research & Development Group, Hitachi, Ltd., Tokyo, Japan
e-mail: chiaki.hirai.xj@hitachi.com

2.1   The Social Issues Japan Faces

The Social Issue Drivers

The problems Japan faces are legion. The country's birthrate will continue to fall, and its population will continue to age. Rural communities are dwindling, and many will decline and become abandoned and desolate. Meanwhile, the population is increasingly concentrated in cities, leading to traffic congestion and a heightened risk of mass-scale damage in a natural disaster. Though cities are supposed to be large population centers, the service-sector jobs therein are increasingly under-staffed. Despite the labor shortage, wages are by no means high, and increasing numbers of young people are in non-regular employment, driving down the birth-rate even further. As the workforce shrinks, so does tax revenue. Nonetheless, gov-ernment spending will continue to rise because of the need to maintain crumbling infrastructure. These factors, coupled with the swelling welfare budget necessary for coping with the graying population, are placing an ever-heavier burden upon the working-age population.

How is Japan to deal with these problems? Rather than addressing the symptoms, it is better in many cases to identify and treat the root causes. Hence, we will distin-guish the social issues themselves from their causal factors. These causal factors can relate to the social issues in very complex, interwoven ways, but if we trace the root causes, we should uncover phenomena that our society, like it or not, will have to acknowledge. It is these underlying phenomena from which social issues derive.

Let us clarify what we mean by "social issue." We define a social issue as a prob-lem in a society that deprives many of that society's members of their lives, prop-erty, freedom, or dignity. We shall call these "social issue drivers" to describe the underlying phenomena that cause these social issues, and which our society must acknowledge, however inconvenient. Social issue drivers are not in themselves problematic. To take the graying population as an example, we could call this trend a social issue driver rather than a social issue, because it can cause the welfare bud-get to swell, which in turn can lead to significant losses in young people's disposal income (their property). In this case, the outcome (young people have less disposal income) is the social issue. What are some of Japan's social issue drivers (the factors that cause social issues)? What issues do these social issue drivers cause?

A Shrinking Labor Pool

Japan's birthrate looks set to continue falling for the time being. This trend has three main effects. First, it leads to an overall population decline and, more importantly, to a decline in the young population—and thus the working population. As Fig. 2.1 shows, the working-age population currently (as of 2015) stands at 76 million, but forecasts indicate that it will dip as low as 52 million by 2050 (Cabinet Office 2017).

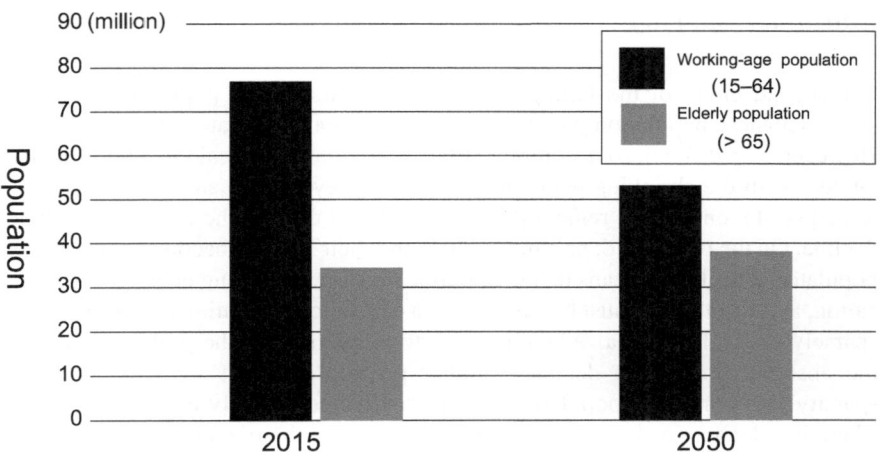

Fig. 2.1 Working population decline (Cabinet Office 2017). Source: Cabinet Office, *Annual report on the aging society*, 2017

A graying population also means that many people of working age will leave the workforce to care for their elderly parents.

Although the labor pool as a whole is shrinking, regional disparities in the labor market have emerged, creating unstable supply and demand. Japan once pursued an economic growth model based on manufacturing. Under this model, both urban and rural areas benefited from economic growth. Manufacturers would establish factories and secure workforces in rural areas, and a transportation infrastructure linking these rural areas of production with urban areas of consumption enabled a reduction in the cost of distributing people and goods. However, the industrial structure has now shifted from manufacturing to services, and many businesses have relocated their factories overseas. This development has deprived rural areas of job opportunities, forcing many young people to move to large population centers. Despite the influx of young workers into cities, insufficient number of workers to prop up the service sector still remains. Convenience stores and transport businesses, for example, are increasingly facing the effects of staff shortages. These understaffed businesses are then criticized for making their staff work long hours and for their failure to maintain service standards.

One possible solution to this problem is to introduce AI and robotic technology. However, unstaffed convenience stores, automated driving, and other forms of radical automatization would result in many jobs being lost. With fewer job opportunities, the young people who came from rural areas would be forced to take low-paying jobs, which offer no prospects for getting married and raising a family. Consequently, the birthrate would plunge even further. Thus, cities absorb rural populations, but they fail to facilitate population increase; consequently, the overall Japanese population continues to decline while also becoming overly concentrated in cities.

Consumer Sparsity

The second effect of the falling birthrate is a sparser rural population. Over the years, Japanese population growth has been accompanied by an expansion of cities. However, it is difficult to accomplish the reverse—to downscale neighborhoods in tandem with the shrinking population. Because they receive an influx of workers, large population centers remain densely populated despite the general population decline. On the other hand, provincial cities and their suburbs become more sparsely populated. This trend means that core public services, including energy, water, education, and healthcare, must be supplied to a consumer population that is distributed sparsely over a large area. When it comes to infrastructure, the problem is not just the absolute population decline; another problem is what we call "consumer sparsity"—a consumer population that is distributed sparsely across a large area (there is a decline in population density). The greater the rate of consumer sparsity is, the higher the infrastructure-related costs per consumer are. When these costs cannot be borne, the quality of the services declines. If there is no adequate water infrastructure, for example, residents might have to head out every day to water supply facilities and haul back water to their homes.

Aging Population

The third effect of the falling birthrate is an aging population. This effect is also related to another causal factor: people are living longer. An aging population means that older people account for an ever-greater share of the overall population, a phenomenon caused by both the falling birthrate and longer lifetimes. With fewer of the population in work, economic growth stalls and national and local governments receive less tax revenue. Nonetheless, an older population entails higher social welfare spending (see Fig. 2.2) (Ministry of Health 2012). With national and local governments in poor fiscal health, citizens must either accept lower quality social welfare or shoulder a heavier burden to maintain social welfare at its current level. Less tax revenue also deprives government of the financial resources necessary to address social inequality or assist vulnerable members of society, resulting in entrenched intergenerational inequality. This situation increases social insecurity, and it robs marginalized people of opportunities by which they could otherwise use their talents. Consequently, Japan will lose its competitiveness, and its productivity will decline further.

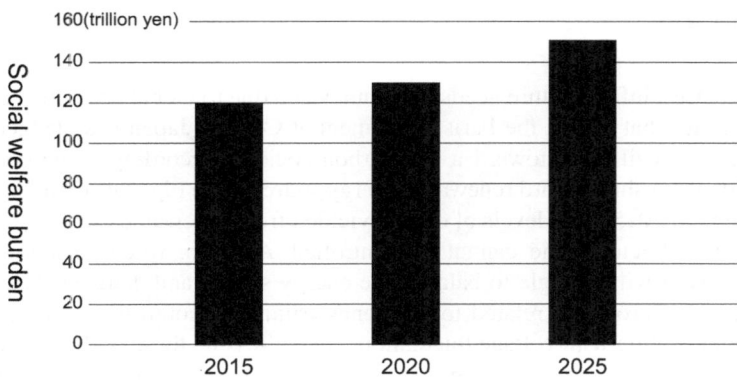

Fig. 2.2 Rising social welfare burden. Source: Ministry of Health, Labour and Welfare, *Shakaihoshō ni kakaru hiyō no shōraisuikei no kaitei ni tsuite (heisei 24 nen 3 gatsu) [Revision to future projection of costs required for social security <March 2012>]*

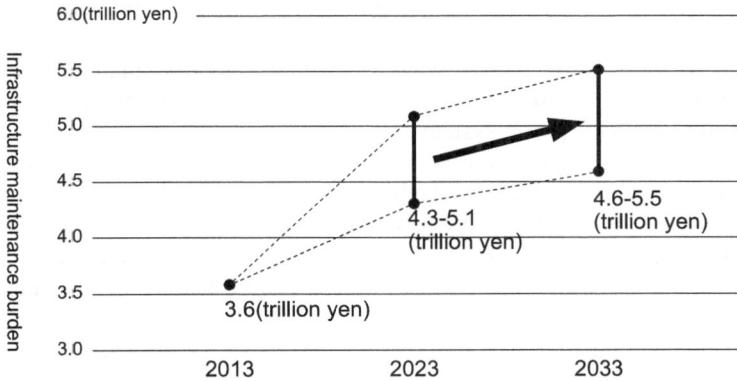

Fig. 2.3 Rising infrastructure maintenance burden. Source: Ministry of Land, Infrastructure and Transport, *ShoraiSuikei [Future projections]*

Aging Infrastructure

As we just learned, consumer sparsity leads to inefficient usage of infrastructure, but crumbling infrastructure is an independent social issue driver besides the demographic problem. Japan's basic infrastructure was developed at a massive scale during the country's high economic growth period, during the 1950s, 1960s, and 1970s. With more than half a century having elapsed since then, Japan's roads, bridges, waterworks, and other infrastructure are decaying, placing upward pressure on social costs (see Fig. 2.3) (Ministry of Land 2013). According to estimates, around 190 trillion yen will need to be provided for infrastructure renewals over a 50-year period from 2011 to 2060.

Shift to Renewables

In some cases, infrastructure needs to be innovated due to global pressure. As one of the countries that signed the Paris Agreement at COP21, Japan has pledged to the world that it will work toward a low-carbon society. Accordingly, the country is committed to a shift toward renewable energy sources (see Fig. 2.4) (Ministry of the Environment 2015). The levels of energy yielded from renewables such as wind and solar power fluctuate and cannot be controlled. As such, when society shifts to renewables, it will struggle to balance the energy supply and demand. The society will also face problems related to frequency trimming, controlling reverse power flows, and dealing with voltage fluctuations. In addressing these problems, the society will need to invest more in power system facilities and raise energy prices to reflect energy yield and storage unit prices. It must also contend with a destabilized power system resulting from the diminished ability to adjust energy demand and supply. Such responses generate new social issues.

Stated differently, to achieve a carbon-free society, Japan must lower renewable energy prices, promote energy saving and more controlled supply and demand in large population centers (where energy consumers are clustered), and provide energy at lower prices to a sparsely distributed consumer population in provincial cities and their suburbs. If Japan fails to deal with these tasks appropriately, energy prices will rise and the power system will become destabilized. These outcomes will then have negative repercussions; as well as causing inconvenience for consumers, they will make businesses less competitive, thus hindering Japan's economic growth and undermining the country's productive capacity.

Figure 2.5 illustrates the above dynamics. There are three relational elements. The first is social issue drivers, which describe unavoidable social trends. These social issue drivers give rise to the second element, which is social issues. Social issues then affect the third element, which is quality of life (QoL).

The next section outlines our view on how we should deal with the social issues.

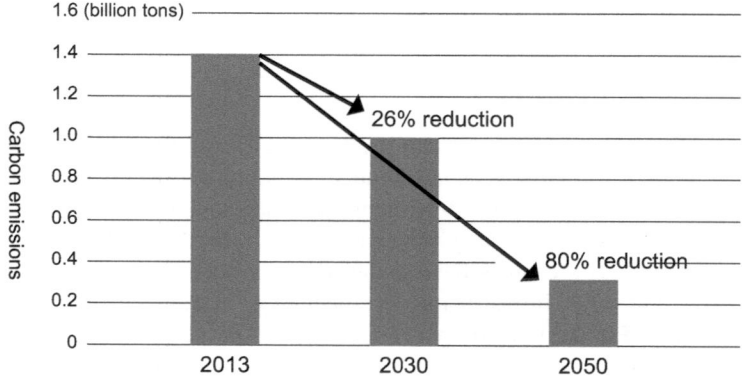

Fig. 2.4 Carbon reduction targets. Source: Ministry of the Environment, *Nihon no yakusokusōan (2020 nenikō no aratanaonshitsukōkagasuhaishutsusakugenmokuhyō [Japan's draft pledge: new targets for reducing greenhouse gases from 2020 onward]*

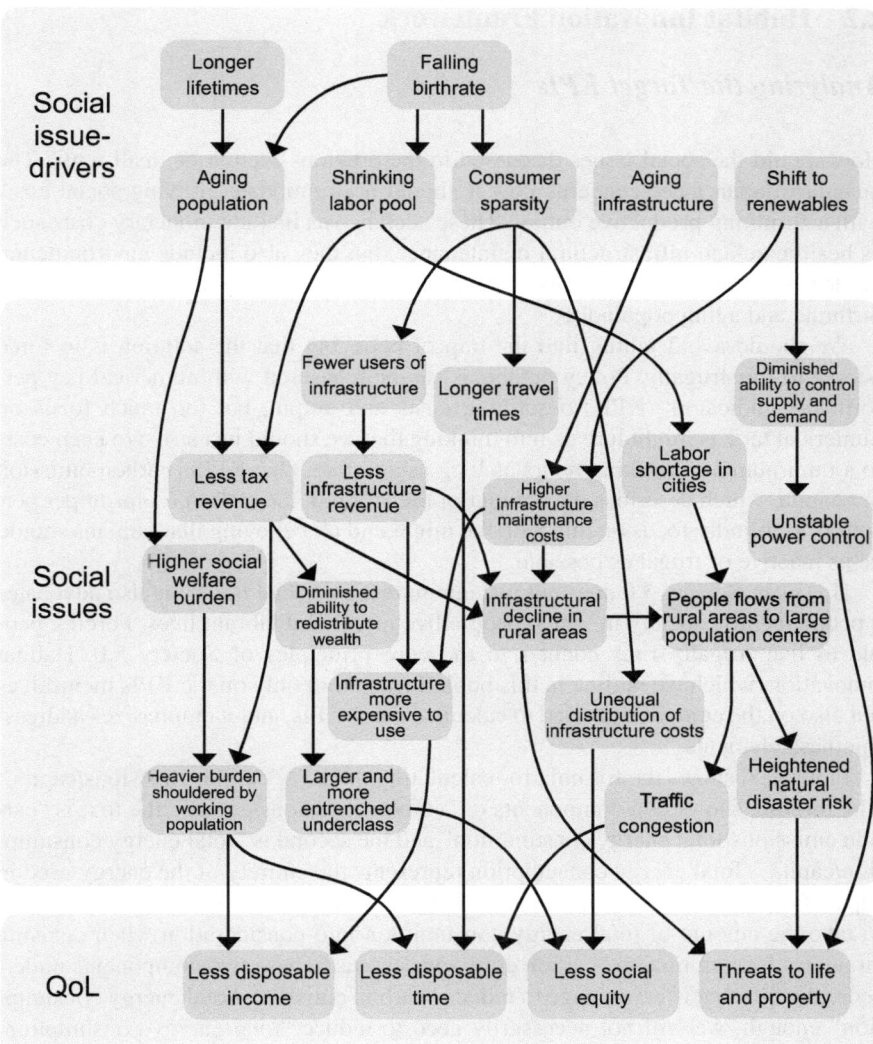

Fig. 2.5 Social issue drivers and social issues

2.2   Habitat Innovation Framework

Analyzing the Target KPIs

How should the social issues discussed in the previous section be dealt with? The quandary Japan faces concerns how it should accommodate growing social costs with a shrinking productive output. These social costs include monetary costs such as healthcare and infrastructural maintenance, but they also include environmental burdens such as carbon emissions. Japan's declining output is driven by its falling birthrate and aging population.

We should avoid falling into the trap of believing that the solution is to force people to live frugally. Policy outcomes can be measured with numerical key performance indicators (KPIs) describing costs and output, but too much focus on numerical targets might lure us into thinking that we should just strive to keep costs to a minimum and working hours as long as possible. One KPI is carbon emission per capita, which describes the amount of reduction in the carbon footprint per person. If we blindly focus on this KPI, we might end up believing that humans should be as inactive or frugal as possible.

However, Society 5.0 does not just aim to resolve social issues; it also advocates a people-centric society in which people live joyful and vibrant lives. Forcing people to live frugally runs counter to the core principles of Society 5.0. Habitat Innovation, which we outline in this book, focuses not only on the KPIs themselves but also on the components used to calculate these KPIs, and it emphasizes addressing these elements.

Figure 2.6 shows the formula for calculating the KPI "carbon emissions/capita." The formula shows two components of "carbon emissions/capita": the first is "carbon emissions/total energy consumption" and the second is "total energy consumption/capita." Total energy consumption represents the entirety of the energy used in Japan. Given that carbon emissions result from energy consumption, it makes sense to take the amount of total energy consumption into consideration when devising strategies for minimizing "carbon emissions/capita." These two components underscore the fact that if we manage to reduce "carbon emissions/total energy consumption" enough, we will not necessarily need to reduce "total energy consumption/capita"; in other words, we will not need to be frugal with our energy consumption.

Fig. 2.6 Analysis of carbon emissions (1)

The only way to achieve the reduction in "carbon emissions/total energy consumption" is to switch to renewables. If we restructure our energy mix, raising the share of renewables (nonfossil fuels such as wind and solar power), then we can consume plenty of energy while still reducing carbon emissions.

However, switching over to renewables for all our energy needs is not feasible in the short term. So does that mean that, for the time being, we limit our energy consumption as much as possible, so as to minimize "total energy consumption/capita?" Let us change the angle slightly and ask this: Is quality of life proportionate to energy consumption in the first place? In asking this question, we are not trying to suggest that we should look to more nonmaterial forms of comfort. Rather, we are suggesting that total activity and energy consumption are, to some extent, independent of each other. For example, if your air conditioner automatically deactivates airflow when the room is vacated, you could save energy without sacrificing any comfort. Similarly, if you replace your light bulbs with LED bulbs, you could cut your energy consumption while enjoying the same light levels as before. Technological progress enables us to maintain the same level of total activity (for example, engaging in work or leisure activities after dark) and still reduce energy consumption (such as by installing LED lighting). With this in mind, we have taken "total energy consumption/capita" in Fig. 2.6 and broken it down into two further components (Fig. 2.7).

Deriving an Approach from the Formula

Figure 2.7 shows three components of "carbon emissions/capita": "carbon emissions/total energy consumption," "total energy consumption/total activity," and "total activity/capita." These are the three factors we analyze in Habitat Innovation as part of our effort to usher in Society 5.0.

The social issue in question concerns the need to minimize "carbon emissions/capita." However, individual members of society wish to increase the third compo-

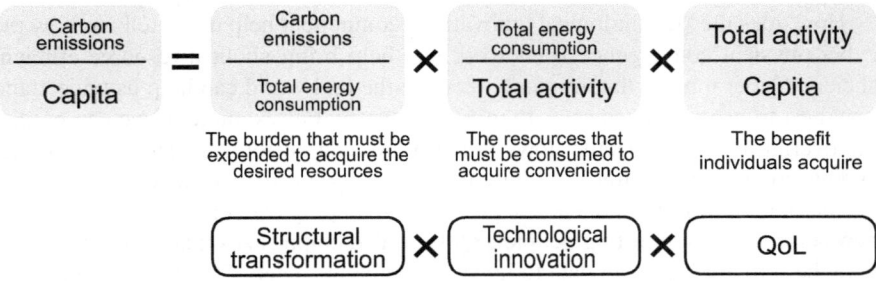

Fig. 2.7 Analysis of carbon emissions (2)

nent—"total activity/capita." To balance these two interests, we must sufficiently reduce the first two of these components ("carbon emissions/total energy consumption" and "total energy consumption/total activity"). To reduce the first component, "carbon emissions/total energy consumption," we must transform the very structure that generates social costs, as in the case of carbon emissions. Let us call this task the "structural transformation." Structural transformation requires government leadership. On the other hand, to reduce the second component, "total energy consumption/total activity," we must find new ways to enjoy a full life—ways that do not require us to use too many resources. To this end, we must look to technology, including automation, optimization, and energy efficiency. Let us call this task "technological innovation." As for the third component, "total activity/capita," this KPI represents our quality of life (QoL). Total activity is in large part conceptual; we do not define it as a numerical metric. In fact, the main point of this formula is to prompt a discussion of how we should define QoL.

Habitat Innovation is not only concerned with environmental issues. Rather, it uses the threefold analytical paradigm (structural transformation, technological innovation, and QoL) to explore how to minimize a whole range of social costs and how to boost productivity.

In Chap. 1, we described Society 5.0 as a data-driven society based on cyber-physical convergence. How does this ICT-infused vision relate to Habitat Innovation's formula? Cyber-physical convergence is ultimately a framework. Habitat Innovation, on the other hand, provides some direction by exploring how we should deploy this framework.

The "structural transformation" component suggests ways in which the cyber-physical convergence framework can be deployed in the policymaking process. To effectuate a massive structural transformation in society, policymakers must analyze quantitative data about the status quo, forecast future trends, and compare potential policy options with existing precedents. In the case of carbon emissions, they must analyze energy demand, forecast long-term energy trends, and evaluate the technological and economic feasibility of renewables. An effective approach for analyzing energy demand would be to visually model societal trends using data gathered from the physical space (real world). An effective approach for the forecasting and evaluation processes would be to run simulations in cyberspace.

How does the "technological innovation" component help us? It tells us how the cyber-physical convergence framework can help bring about a resource-efficient society. Cyber models that minutely recreate the real world can help us understand how best to use resources so as to minimize waste. For example, if we can predict the overall pattern of people flows in a city, we can customize transport, lighting, and air-conditioning patterns to these movements so as to avoid waste.

As for QoL, this component can prompt us to deploy data in a way that generates new services for supporting people's QoL. It also emphasizes that residents should take the initiative in using data to make a change in society. When data are used to gain quantitative insights into social issues, they allow all the stakeholders, including government, businesses, and residents, to share their views and discuss the issue on a level playing field. In this sense, data are a crucial tool for encouraging residents to come forward and communicate their interests and concerns with government, businesses, and each other.

Residents as the Actors of Innovation

There are existing cases of structural transformation and technological innovation in practice, and these have been led by government and businesses. Government has been the instigator of structural transformation, while businesses have been the actors of technological innovation. However, the critical impact of the data-driven society is the potential for residents to use data and become the chief actors of innovation. It was with this in mind that we named our approach Habitat Innovation.

However, individual residents will only respond to this opportunity once the practice of resident-led data gains traction. Additionally, the public must use the data effectively. Otherwise, the stability of society might be threatened. For example, data might be leveraged by some residents for their own personal ends, or the government might become wary of letting the public take the lead if public opinion is too easily swayed by a short-term outlook. We must therefore ask the question: How can we ensure that the public, government, and businesses are sufficiently literate for the digital age?

Government and businesses must use reliable data and become more open. The public, for their part, must engage with government and businesses continually and proactively while generating data themselves, and government and businesses must duly respond to them. To ensure that this cycle leads to a more data-literate society, it is essential to progressively develop best practices and foster a conducive culture. Once there is a critical mass of stakeholder consultations over services, technologies, and laws, the public will increasingly become the chief actor of society, and innovation will be increasingly instigated by and for the public. This is what Habitat Innovation is all about. According to Habitat Innovation, once the public take the initiative in using data, it will be possible to balance the resolution of social issues with economic growth and create the conditions necessary for sustainably transforming cities; this is how Habitat Innovation can help usher in Society 5.0.

In Habitat Innovation, the insights of engineering, social sciences, humanities, and many other disciplines are used to analyze what QoL means at an individual level and to identify the role that policy and technology should play in enhancing it. Habitat Innovation further proposes that alongside this, we should develop platforms that enable interdisciplinary data sharing, technologies that can simulate the benefits for society and individuals, and a system architecture that precisely tracks and responds to long-term demand fluctuations and latent needs. These developments must be practically applied in a manner that puts the public first, so as to achieve a sustainable Society 5.0.

The next section analyzes several social issue drivers through the lens of Habitat Innovation. In each case, the more one contributes to structural transformation and technological innovation components, the more the QoL component is improved.

The question of how to improve QoL, which includes the matter of how to define it, is discussed throughout this book; for now, we will discuss strategies for improving the other two factors: structural transformation and technological innovation.

2.3 Using the Habitat Innovation Framework to Solve Key Social Issues

Shift to Renewables

As a means of achieving a carbon-free Japan, the task of shifting to renewable energy entails many social issues. In shifting to renewables, Japan must develop the necessary transmission and distribution infrastructure. This infrastructural requirement will force up energy prices. Energy prices will be further increased by another factor: energy still costs more to generate from renewable sources than it does from conventional sources. Another issue concerns the difficulty of adjusting energy demand and supply; as energy yields vary depending on the weather, it is not always possible to generate energy on demand. Imbalance in supply and demand incurs the risk of major power outages.

Habitat Innovation aims to reduce carbon emissions while maintaining happy and comfortable lives. The formula in Fig. 2.7 provides a working framework for balancing comfort and happiness with carbon emission reduction. The component on the far left, "carbon emissions/capita," represents the amount of carbon dioxide emitted per person. To minimize this metric would be in society's best interest. "Carbon emissions/capita" is the result of multiplying the three components on the right with each other ("carbon emissions/total energy consumption," "total energy consumption/total activity," and "total activity/capita"). Because the principle of Society 5.0 is to maintain or maximize QoL (the third component in our analytical paradigm), we can only reduce the other two components.

Structural Transformation

In this case, structural transformation underscores the importance of reducing the amount of carbon dioxide we produce when consuming a given amount of energy. In other words, it implies that we must increase renewable energy as a share of the total energy we use. In doing so, we must fulfill three requirements: our energy must be ecologically sustainable, stable, and economically viable. Accordingly, when considering the basic power system, for instance, we must ensure that local power distribution networks are able to deliver power to consumption zones within the region, and that the broader power transmission networks can deliver power across each region. To ensure the stability of the power system, it will be necessary to deploy IT-based and finely tuned control technologies. To forecast demand more effectively, it will be necessary to gather data from stakeholders, including businesses and social actors. Policy recommendations will be essential in encouraging investment in facilities and innovation, as well as in facilitating the gathering and deployment of data.

Technological Innovation

In this case, technological innovation is a means to eliminate wasteful energy consumption at a society-wide level. The aim is to encourage society to use energy less wastefully, so as to bring down the total energy consumption without having to effectively curb people's activity. In public transport, for instance, if timetables were regulated dynamically to reflect ridership levels, it would help curb wasteful energy use. Another example is courier services, where much energy is spent on redeliveries (due to the absence of the recipient). Technology could help couriers cut out wasteful energy use by monitoring whether recipients are at home and by plotting the most efficient routes. Of course, these technologies must be accompanied by technologies that ensure personal privacy. Industrial actors and large building operators could combine data gathered from IoT devices and sensors to derive precise estimates, from which they could run simulations so as to minimize peaks in energy use or share energy with other regions. Such action will optimize the overall supply of energy; moreover, when we do switch over to renewable energy, this action will have enabled us to prevent a wasteful swelling of the energy supply facilities. It will be good news for consumers too: electricity rates will be lower. Technological innovation also has a crucial role to play when it becomes necessary for consumers to change their habits. That is, technology can help consumers pursue their desired activities in more flexible ways and ultimately guide them into more energy-efficient behavior. To enable such technological innovation, we first need to develop a data networking infrastructure spanning social actors and industries.

The Shrinking Labor Pool

Japan's labor pool is shrinking because of its dwindling population, itself a result of the falling birthrate. Japan currently relies on foreign workers to plug the labor shortfall. In cities, many foreigners work in the service sector; in rural areas, many are propping up farming businesses. But this strategy might not be sustainable in the long term. As populations across Asia get older, Asian nations will increasingly compete over foreign workers, which may lead to lower numbers of foreign workers in Japan. Eventually, Japan's foreign workforce will start declining alongside the indigenous workforce. The labor problem is also related to productivity. Workers in Japan tend to be less productive than their counterparts in other developed nations, and they make up for this with the deep-seated practice of long working hours. Japan has some room for improvement in this regard.

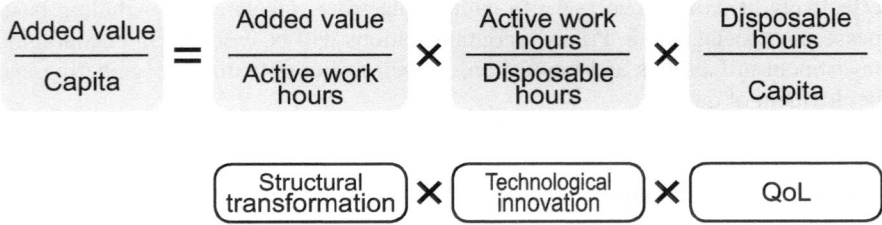

Fig. 2.8 Analysis of labor productivity

Habitat Innovation has broken down the issue into the formula shown in Fig. 2.8. On the left is "added value/capita," which represents how much added value each person produces. A society must try to maximize this metric. The KPI is the result of multiplying the three factors (structural transformation, technological innovation, and QoL) together, which are shown on the right-hand side of the figure. In this case, structural transformation represents the added value produced per working hour, technological innovation represents the number of hours that must be spent in work for each disposable hour, and QoL represents the number of disposable hours allotted to each person. In other words, structural transformation relates to productivity, technological relates to work time, and QoL relates to free time. Habitat Innovation aims to raise productivity sufficiently to allow a reduction in work time and an increase in free time. As important as work may be, a vibrant life also requires plenty of free (or "disposable") time.

To calculate labor productivity, it helps to divide work time into net "active work hours," meaning the time workers spend in value-generating work, and "waiting hours," when workers are commuting or traveling as part of their job or are waiting for the resources to be prepared. If you increase productivity per active work hour and decrease waiting hours, you will generate more disposable hours (free time). Readers ought to note that we have defined travel time and the like as "waiting hours" rather than "work hours" for the sake of convenience; whether such time is legally deemed to be work time is another matter.

Structural Transformation

In the context of this issue, structural transformation underscores the need to raise the amount of value produced per active work hour—in other words, the need to boost value productivity. Finding a place to work is an essential requirement for a vibrant life. The objective of data deployment and AI-driven automation is not to take away job opportunities but to create new industries. Historically, automation has transformed livelihoods and jobs, but it has always generated new industries too. With Habitat Innovation, industry-spanning data are combined in such a way as to yield analytical tools that can help identify new needs and business opportunities. To this end, businesses must be incentivized to move toward value-added services, and there must be a legislative infrastructure to encourage more open data, a well-spring of innovation. Another task is to offer proposals for education policies aimed at nurturing the ideal workforce—individuals with the creative ideas and practical know-how necessary to produce new value, and who can adopt work styles fit for the digital age.

Technological Innovation

Technology has a major role to play in reducing waiting hours and increasing disposable hours. For example, VR and communications technologies can cut travel time by recreating the office environment in workers' homes or satellite offices. Similarly, where once it was in the realm of science fiction, it is now becoming possible for people to port themselves into a robotic avatar and exist virtually in a remote worksite. Even when workers have to travel physically, greater travel efficiency can be realized by mobility-as-a-service (MaaS), which describes a trend to combine transportation services dynamically so as to provide the most efficient travel possible. Using data gathered from the real world, businesses can coordinate their schedules with each other to minimize lag time that would result from unsynchronized schedules.

Aging Infrastructure and Consumer Sparsity

Aging infrastructure and consumer sparsity are closely related, so we shall discuss them together. Much of Japan's infrastructure was erected during the country's period of high economic growth, and it is rapidly approaching its expiration date. To maintain this infrastructure, renewal is necessary. However, users are dwindling due to depopulation, and so there is no need to maintain the infrastructure at its present scale. Accordingly, infrastructure should be renewed, but in a manner that reflects the diminished population.

It is important to realize that we cannot simply downsize infrastructure in proportion to the population. When Japan's population was growing, neighborhoods expanded in tandem with the growth. However, in this era of population decline, living space remains as expansive as before, while the population grows ever sparser (i.e., while the density decreases). Accordingly, if service provision is scaled down in proportion to the rate of population decline, many users would be inconvenienced. Suppose that the number of elementary schools in a given area is reduced to reflect the dwindling number of children. Because the residential spaces remain as expansive as before, many families will have to endure a longer school commute. The same goes for hospitals and retail stores; when these facilities are reduced to reflect the shrinking demand, many users will have to travel further to the few that remain. Likewise, the services that deliver goods, water, and energy will be supplying these services to fewer users but over an equally large area, which means a heavier cost of service delivery per user.

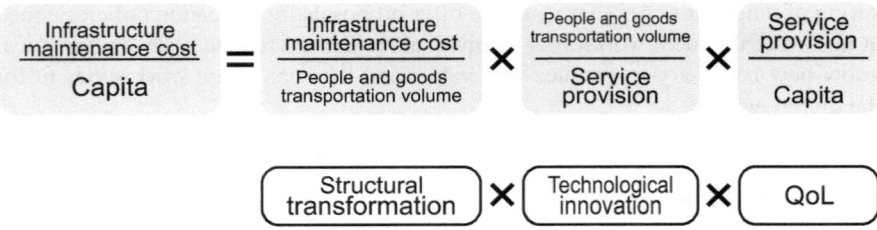

Fig. 2.9 Analysis of aging infrastructure

Consumer sparsity is associated with greater costs in terms of time and money. The problem could be solved by making communities more compact, but until then, we must find a way to ensure that sparsely distributed consumers are not inconvenienced.

Habitat Innovation has broken down the issue of aging infrastructure into the formula shown in Fig. 2.9. On the left is "infrastructure maintenance cost/capita." Reducing this metric is in the best interest of society. On the right, the third component (QoL) is "service provision/capita." To increase this metric (while decreasing "infrastructure maintenance cost/capita") would align with the ethos of Society 5.0. Accordingly, we must minimize the other two components: "infrastructure maintenance cost/people and goods transportation volume" (which corresponds to structural transformation) and "people and goods transportation volume/service provision" (which corresponds to technological innovation).

To minimize the first of these two components is to reduce infrastructure maintenance costs while retaining the same level of people and goods transportation. An effective method for this is to make neighborhoods smaller. If communities are more compact, this would reduce transport distances, and we could then achieve a general decrease in infrastructure maintenance costs without needing to reduce the volume of goods delivered. However, this goal is not feasible in the short term. A more short-term measure is to reduce the second component, "people and goods transportation volume/service provision" (technological innovation). The aim here is to leverage technology to reduce physical transport volume while retaining the same level of service provision.

Structural Transformation

In this case, structural transformation means downsizing social infrastructure assets—in other words, making neighborhoods more compact. If expanses of suburban neighborhoods were reorganized into more compactly distributed neighborhoods, the infrastructure in these neighborhoods could be accessed by many more people than before. The problem, however, is that forcing people to move would run counter to the people-centric ethos of Society 5.0, as it would rob people of the ability to exercise choice in their pursuit of happiness. We must therefore make urban life attractive enough that residents naturally want to move into compact cities. Then, as the environs become more depopulated, the configuration of the social infrastructure there should be scaled back flexibly, in tandem with the depopulation rate. To take transport as an example, there should be a rail network within the city, but on the outskirts, people could use buses or car-sharing schemes. To this end, there needs to be the necessary regulatory easing and urban planning.

One effective strategy for downsizing infrastructure is to focus on minimizing peak demand. In the case of energy, for instance, storage batteries could help control peak demand, meaning that you could use smaller power generators. When the facilities are smaller, it is much simpler to phase them out. Another way to control peak demand is to enlist the cooperation of residents. For example, residents could be called on to avoid morning and evening rushes. There should be no coercion though; the infrastructure should be designed in such a way as to incentivize cooperation, and this cooperation should not compromise convenience and comfort.

Technological Innovation

In this case, technological innovation means developing technology that enables infrastructure to operate in sparsely populated expanses at minimal cost. Transport services, for instance, could optimally supply demand with a combination of bus services and car sharing, which can enable dynamically variable scheduling. The running costs could be controlled by limiting unnecessary services and maximizing ridership. On the other hand, IT applications can deliver services without necessitating travel. Examples include remote learning, remote healthcare, and remote elderly care monitoring, each of which can be operated at a cost lower than the cost of traveling for such services. Automated driving and drones can reduce the personnel expenses for goods deliveries. Technology that monitors whether recipients are at home can be used to plan delivery routes and cut down redeliveries. In this way, we must use data, IT, and robotics to lower the costs of public services.

Society 5.0 balances the best interests of society as a whole (resolution of social issues) with the best interests of individuals (people-centric society). In this chapter, we discussed the KPI formula as an approach for balancing these two concerns. Habitat Innovation proposes using this approach to address social issues. The measures under such an approach are discussed in detail in Chap. 4 onward. The next chapter, however, focuses on a precursor to Society 5.0, the smart city. The chapter discusses the smart city initiative and the challenges it has encountered.

References

Cabinet Office (2017) Annual report on the aging society. https://www8.cao.go.jp/kourei/english/annualreport/2017/2017pdf_e.html. Accessed 4 Jun 2019
Ministry of Health, Labour and Welfare (2012) Shakaihoshōnikakawaruhiyō no shōraisuikei no kaiteinitsuite (heisei 24 nen 3 gatsu) (Revision to future projection of costs required for social security, March 2012)

Ministry of Land, Infrastructure and Transport (2013) Shoraisuikei (Future projections) (infrastructure maintenance information website). http://www.mlit.go.jp/sogoseisaku/maintenance/_pdf/research01_02_pdf03.pdf. Accessed 4 Jun 2019

Ministry of the Environment (2015) Nihon no yakusokusōan (2020 nenikō no aratanaonshitsukō kagasuhaishutsusakugenmokuhyō) (Japan's draft pledge: new targets for reducing greenhouse gases from 2020 onward), July 2015. https://www.env.go.jp/earth/ondanka/ghg/2020.html. Accessed 4 Jun 2019

Chapter 3
From Smart City to Society 5.0

Atsushi Deguchi

Abstract This chapter overviews the history of smart city and smart community projects implemented in Japanese cities since Japanese national government had initiative through the subsidies and supports for the pilot projects promoted by municipalities following the Kyoto Protocol. It reviews that the original technologies of smart grids, microgrids, and smart house, which were created by integrating IT with the energy management system, have been implemented into the pilot projects of the first generation of smart community in 2000s under the condition that Japan has lagged behind in the electricity market liberalization compared with the EU and the USA.

Sections 3.3 and 3.4 review the social background and the process for national government to promote the pilot projects and the energy-conscious policymaking in the local cities in Japan. It summarizes the achievements of the first-generation pilot projects for constructing CEMS-based smart community in 2000s, and characterizes the next-generation smart city models based on the energy management system implemented in 2010s with initiative by private sectors. Section 3.4 explains that Japanese national government has had initiative to activate not only the pilot projects but also policymaking in municipalities following the concept of sustainable urban development and the SDGs.

Section 3.5 characterizes the trends of smart city in Japanese cases with the comparison of cases of the EU and the USA. It suggests on the directionality of the future smart city shifting from the top-down type with initiative by government or big companies to bottom-up type with citizen-oriented technology based on the concept of "Society 5.0."

Keywords Community Energy Management System (CEMS) · Connected community · Open data · Sensing city · Sustainable city

The original version of this chapter was revised: This book was inadvertently published with the incorrect license type CC BY 4.0 and the Open Access License has been amended throughout the book to the correct license type CC-BY-NC-ND. The correction to this chapter is available at https://doi.org/10.1007/978-981-15-2989-4_9

A. Deguchi (✉)
Department of Socio-Cultural Environmental Studies, Graduate School of Frontier Sciences, The University of Tokyo, Tokyo, Japan
e-mail: deguchi@edu.k.u-tokyo.ac.jp

3.1 What Is a Smart City?

Integrating IT into Urban Planning to Smartify Cities

So far, we have outlined the ideas related to Society 5.0. In this chapter, we look back at attempts to implement a smart city, a concept that involves integrating IT with urban planning. Society 5.0 aims to take us beyond the smart city to the supersmart society, but for now we will explore where and how the smart city has advanced. We will also consider how the smart city concept relates to Society 5.0. Various smart city initiatives have been implemented across the world (Nikkei BP Clean Tech Institute et al. 2011). Here, we look back at the smart city initiatives that have been conducted in Japan and Western countries since the turn of the century and examine how far they have come.

There are now countless examples of initiatives that integrate IT into urban community services or that use IT to enhance services or develop new businesses. Such initiatives are particularly numerous in the transport and energy sectors. Many of Japan's bus services, for instance, now use geo-positioning technology so that passengers can tell where the bus is and how long it will take to arrive at the bus stop. Within cyberspace, the bus's spatial information is progressively plotted out along with its movement, and the latest information is relayed to the smartphones of people waiting at the bus stop. Thanks to this information, the people at the bus stop can tell how near the bus is, just as people waiting for an elevator can see which floor the elevator is currently at. People can then mentally process this information, which makes the waiting process much less irritating than if they had no such information and thus no idea of how much longer they have to wait. Another example is automotive navigation systems that use digital mapping in cyberspace. Vehicles monitor or predict the conditions further down the road and relay the information to the driver so as to guide the driving. At an even more advanced level, automated driving is now ready for practical application.

These navigation systems help resolve or avoid traffic congestion and, in so doing, minimize the time and energy of travel. This does not just apply to cars and buses; IT integration will help solve congestion issues for many different transport services, including taxi and rail transport. With such integration, the existing systems underlying community services will be regenerated as highly intelligent, or "smart," systems. Once services are powered by such smart systems, the urban community as a whole becomes a smart city. In Society 5.0, these smart systems will be even more advanced. They not only make life more convenient and comfortable for each city dweller but also help to resolve issues affecting the population as a whole, such as global warming and aging of the population.

The idea that integrating IT with existing services will lead to a more advanced society, as well as the notion that societies advance in tandem with technological evolution (including progress in IT), is a key assumption in the vision of Society 5.0 as outlined in the government's Science and Technology Basic Plans. This literature mentions the "supersmart society," describing it as a society in which cyberspace is proactively used to successively create new value and services that enrich the lives of the society's members. The government's vision of Society 5.0 is shared by the Japan Business Federation (Keidanren). In *Revitalizing Japan by Realizing Society 5.0* (Japanese Business Federation 2017), Keidanren states that Society 5.0 goes beyond optimization of individual fields to the optimization of society as a whole by freeing people from spatial and temporal constraints, freeing them from complex social issues, and encouraging economic growth underpinned by new business models and worldwide proliferation of such models.

Common Urban Infrastructure: From Test Bed to Practical Application

One example of a new system that has resulted from integrating IT with existing services is smart energy. Following the turn of the century, smart city initiatives rapidly spread around the world. Although these initiatives initially focused on introducing new energy systems, they helped make the smart city concept more widely known, including in Japan.

Nowadays, the smart city initiatives of local governments and private businesses go beyond energy to encompass a breadth of community services, including those related to transport, healthcare, welfare, and waste disposal, such that the concept itself is now much broader. Many cities in Western and Asian countries have piloted and rolled out the smart city model, making the concept even more far-reaching.

As the name suggests, the smart city denotes an "intelligent" city. Smart cities integrate IT with various services, activities, and physical things (energy and rail systems being examples) to improve convenience, comfort, and safety in the city. Smart cities also integrate IT with services to address the issues the city is facing. Cities face their own particular problems; they also face problems that are common across society. Japan, for one, faces a myriad of problems. Suffice it to say, the solutions to these problems must take into account the particular conditions of the city in question, including its social and geographical conditions. Some strategies are applicable to multiple cities, but some measures are only relevant to certain cities.

Likewise, basic infrastructure, such as power systems, is applicable to multiple cities, and so the technology underlying the systems can be shared among them. That said, there are some discrepancies between countries. Western countries are ahead of Japan when it comes to solar and wind power and the liberalization of energy markets. These countries have pressed ahead in introducing smart grids and similar systems in an effort to diversify energy sources and accommodate diverse user preferences. The following section outlines some examples of smart city initiatives that have been implemented around the world since the turn of the century.

3.2   Smart Energy Management Systems

Smart Energy Supply Systems

One commonality of cities is that they rely on energy systems. It is no surprise that many smart city initiatives have focused on "smart" energy—energy systems that integrate IT with local power supply systems. The main roles of smart energy systems are played by smart grids, microgrids, and smart houses.

Smart grids, microgrids, and smart houses were the leading players in the first phase of smart city initiatives, and they have therefore become key terms associated with the smart city concept. Applications of "smart grid" and "microgrid" technologies have primarily involved the construction of an advanced energy management system, which in Japan is called "Community Energy Management System (CEMS)." The term "smart houses," on the other hand, is associated with rows of detached houses that use a Home Energy Management System (HEMS) to achieve optimal energy management.

The practical application and proactive rollout of these new forms of energy management have been critical in advancing the smart city concept. In the next section, we explore each in more detail.

Smart Grids

A smart grid is an electrical grid that applies IT to power supply facilities so as to optimize energy supply. For example, it links energy supply and demand in an information network; it introduces control mechanisms that would be unfeasible under conventional, centrally controlled energy supplies; it cuts costs by enabling compatibility with diverse energy sources and by optimizing the supply–demand balance within the transmission network; and it controls load bearing so as to prevent power outages. Japan's power grid has, over the years, been based upon a centralized power control system, in which the country's power infrastructure is divided into several territories, each controlled by a major power company. This system has ensured a stable power supply. By contrast, the USA has liberalized energy markets. One problem with this fragmented power distribution system is infrastructure maintenance. As the infrastructure ages, major power outages increasingly occur. The USA has rolled out smart grids partly as a response to this problem. American smart grids use smart meters, which monitor and communicate information, and thus enable a greater level of control than that achievable in conventional grids. The greater level of control includes the ability to avoid excess load bearing or accidents in fragile transmissions as well as the ability to modify transmission routes. As for the EU, which has made headway in introducing renewable energy sources (such as solar and wind power), its smart grids integrate IT with grids (transmission networks) to enable compatibility with diverse energy sources and to optimize the supply–demand balance within the grid (transmission network).

Microgrids

Whereas smart grids are wide-area (macro) grids, microgrids are localized grids. Microgrids source energy from renewables such as solar, wind, and biomass, and use IT to monitor and control the supply. These grids do not rely on large power stations and thus avoid the problems associated with them, which include environmental problems and energy losses incurred from transmitting the power to remote locations. The energy in microgrids is both sourced and consumed locally. Hitachi has the following to say about the US microgrid market:

> Most of the time, microgrids operate in parallel with the utility grid but also have the unique feature of being able to operate independently of the main utility grid (island mode) in the event of a power outage. As dependence on technology has grown in all facets of society, tolerance for power outages has decreased markedly while at the same time in the USA, vulnerability to power outages has increased due to aging of the grid infrastructure and cyber and physical threats. This makes the ability to seamlessly "island" from the utility grid in the event of a power outage a key driver for many customers to consider a microgrid versus other less sophisticated distributed energy resource (DER) solutions. Additional customer benefits include reduced energy costs, less volatile energy costs, and reduced emissions.
>
> Microgrids strengthen energy stability and the ability to recover from outages. They also reduce the carbon footprint and, in many cases, reduce overall energy costs. The benefits to the public are formidable. All across America, there are efforts to increase communities' ability to recover from natural disasters, terrorist attacks, and other threats to national security (Aram 2017a, b).

Smart Houses

Smart houses are houses that connect appliances and equipment to communications lines to achieve optimal control. This concept was proposed back in the 1980s, but with the advent of the Internet and digital home appliances, as well as the spread of broadband, the concept was extended to control systems that use the home's Internet connection and systems for monitoring elderly people and children. Following the recent rollout of HEMS, there has been a rise in the number of smart houses equipped with unitary control systems—systems that coordinate all energy supply and consumption by integrating home appliances, solar power systems, batteries, and electric cars.

Many of the smart communities and smart cities presented in the next section have introduced the aforementioned CEMS and built smart house models that reflect the attributes of the community concerned. In the following section, we explore taxonomies and trends as they relate to Japan in a little more detail.

3.3   Japan's Smart Communities/Cities

Smart Communities That Use Community Energy Management Systems

As Western countries continue to roll out advanced energy supply systems such as smart grids and microgrids, Japan, which lags behind in energy liberalization, has started piloting such systems in certain communities. In these communities, local governments work with private businesses in implementing state-subsidized projects designed to reduce greenhouse emissions and contribute toward a carbon-free society.

These projects are anchored within government strategies such as the Kyoto Protocol Target Achievement Plan, which the Cabinet formulated in April 2005. The Kyoto Protocol Target Achievement Plan was designed to meet the 6% reduction in greenhouse gas emissions, to which Japan committed when it signed the Kyoto Protocol agreed to at the Conference of Parties III (COP3) in 1997 (the 6% reduction is relative to the 1990 level; reduction in hydrofluorocarbons is relative to the 1995 level). Under this plan (which was subsequently revised in March 2008), relevant government departments adopted a system for supporting national projects such as a test bed pilot program, which has spurred considerable action.

In November 2009, the Ministry of Economy, Trade and Industry (METI) launched the Council for Next-Generation Energy and Social Systems and developed a test bed pilot program titled the "Next-Generation Energy and Social Systems Demonstration" (Ministry of Economy 2019). This program outlined five objectives: (1) "Stable accommodation of large-scale roll-outs of renewable energy" (develop a robust power infrastructure that can handle large-scale expansion of renewable energy); (2) "IT-driven optimization and load distribution" (showcase next-generation lifestyles that use IT to balance QoL with energy saving); (3) "A growth strategy that markets the system" (showcase the system overseas as part of a growth strategy); (4) "Standardization" (lead the world in establishing next-generation international standards); and (5) "A business environment that will take the technology from testbed to practical application" (develop a financing system involving collaboration with relevant government departments [e.g., Ministry of Land, Infrastructure, Transport and Tourism; Ministry of Agriculture, Forestry and Fisheries; Ministry of Education, Culture, Sports, Science and Technology] and develop the frameworks for autonomous financing; review relevant systems). In April 2010, four municipalities were designated as test bed sites: Yokohama (Kanagawa Prefecture), Toyota (Aichi Prefecture), Keihanna Science City (Kyoto Prefecture), and Kitakyushu (Fukuoka Prefecture) (Tsuchiya 2015).

Over the next 5 years, these municipalities pursued their own projects with residents' participation and in collaboration with private businesses. Yokohama's project was titled Yokohama Smart Community, Keihanna Science City's (aka Kansai Science City) was titled Keihanna Eco-City Next-Generation Energy and Social System, Toyota's was called Smart Mobility and Energy Life in Toyota City ("Smart Melit"), and Kitakyushu's was called Kitakyushu Smart Community Creation Project (Ikeda and O'oka 2014; Architectural Institute of Japan 2014).

The aim of each project was to construct a CEMS in the existing urban areas and evaluate how effectively it operated, taking into account local attributes. The projects introduced technologies such as HEMS, Building Energy Management Systems (BEMS), storage batteries, and electric vehicles into houses and buildings in the existing urban areas. The aim was to integrate these systems with demand response and incentive point schemes so as to construct an organic CEMS. The rise of the CEMS is one of the products of Japan's efforts to build smart communities and cities.

In addition to these test bed programs, METI launched a public appeal for related projects and selected such projects as "smart community vision proliferation," "next-generation energy technology test bed," "smart community rollout facilitation," and "smart energy rollout facilitation." These projects are implemented by local governments, private businesses, or both collaboratively, at a nationwide scale.

The Smart City Concept in Large Urban Development Projects

Some urban development projects involve building upon the existing infrastructure. In other words, the city has its existing assets, such as the energy supply systems, core urban infrastructure, housing, and the like, and new systems are added to the neighborhoods situated among these facilities. By contrast, in large urban development projects, the city is built from scratch. This makes it much easier to introduce cutting-edge physical infrastructure, including that related to power transmission lines, roads, gas supply, and communications. Another advantage is that the high-tech infrastructure will enable new neighborhoods and innovative lifestyles, allowing the municipality to brand itself as a "new town."

During the 2010s, there have been a number of smart city projects in Japan's new towns. The key examples include Kashiwa-no-ha Smart City (Kashiwa, Chiba Prefecture) (Yamamura 2015; Mitsui Fudosan 2019) and Fujisawa Sustainable Smart Town (Fujisawa, Kanagawa Prefecture) (Fujisawa Sustainable Smart Town Association 2019). Kashiwa-no-ha Smart City is situated around Kashiwa-no-ha Campus Station (which is served by the Tsukuba Express) in a 273-hectare plot designated as a land readjustment project area (planned population: 26,000). During the 2011 Tohoku disaster, the smart city underwent a planned power outage. In the same year, the national government designated Kashiwa city as one of "FutureCities" (more on this in Sect. 3.4), making the new town eligible for government subsidies. As an eco-model city, Kashiwa works with Mitsui Fudosan in pursuit of three objectives: eco-friendly urban development, longer healthy life expectancy, and creation of new industries—including in new, economically invigorating growth sectors.

For the first objective (eco-friendly urban development), an "Area Energy Management System (AEMS)" was introduced to manage energy supply in four zones around the station (a mixed-zone housing commercial facilities, hotels, and offices; a zone housing large commercial facilities; and two zones housing high-rise apartment buildings). This system was developed by Hitachi. The town also introduced a business continuity plan for emergencies, which includes the use of large storage batteries and a gas-fired power generator, and it provided energy interchange between the different zones intersected by roads (see Figs. 3.1 and 3.2). Initiatives for the second objective (longer healthy life expectancy) include the opening and running of a healthcare station called Ashita ("tomorrow"). Initiatives for the third objective (creation of new industries) include the opening of the Kashiwa-no-ha Open Innovation Lab and the Kashiwa-no-ha IoT Business Co-Creation Lab.

Of note here is that these urban development initiatives have been coordinated by the Urban Design Center of Kashiwa-no-ha (UDCK, founded in 2006), a platform for government, business, and academic collaboration. Government–business–academic collaboration proved instrumental in developing the AEMS. It helped get Kashiwa-no-ha Smart City designated as both a "FutureCity" and a "special zone for local economic invigoration," which made the smart city eligible to apply for a special zone status. With special zone status, the smart city

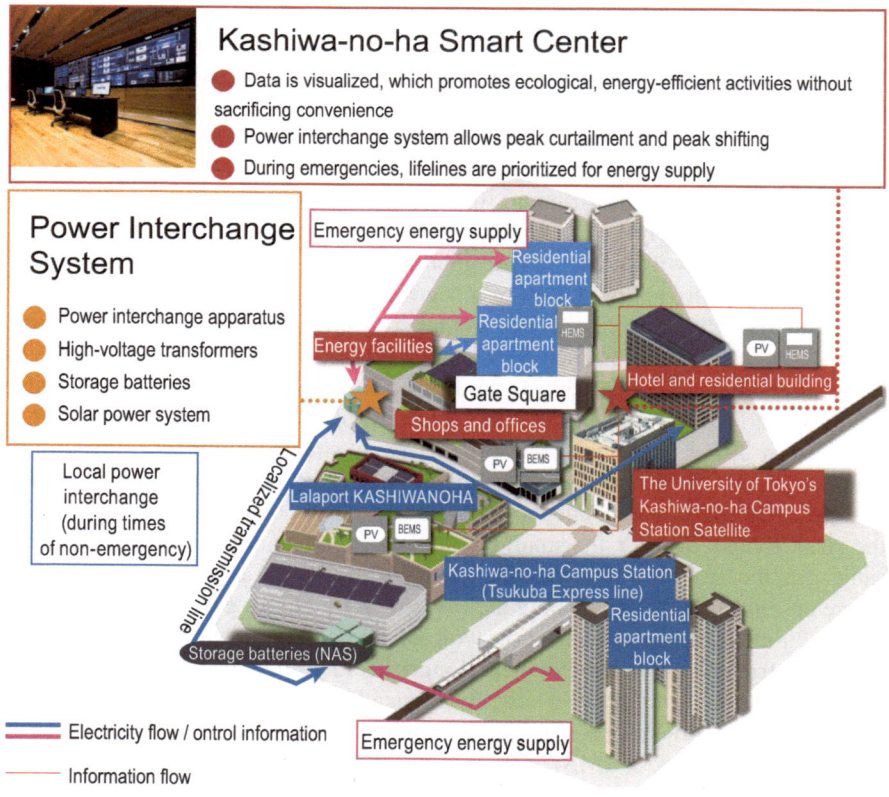

Fig. 3.1 Urban redevelopment area, including Kashiwa-no-ha Smart City. Source: Mitsui Fudosan, *Kashiwa-no-ha Smart City* (Mitsui Fudosan 2019)

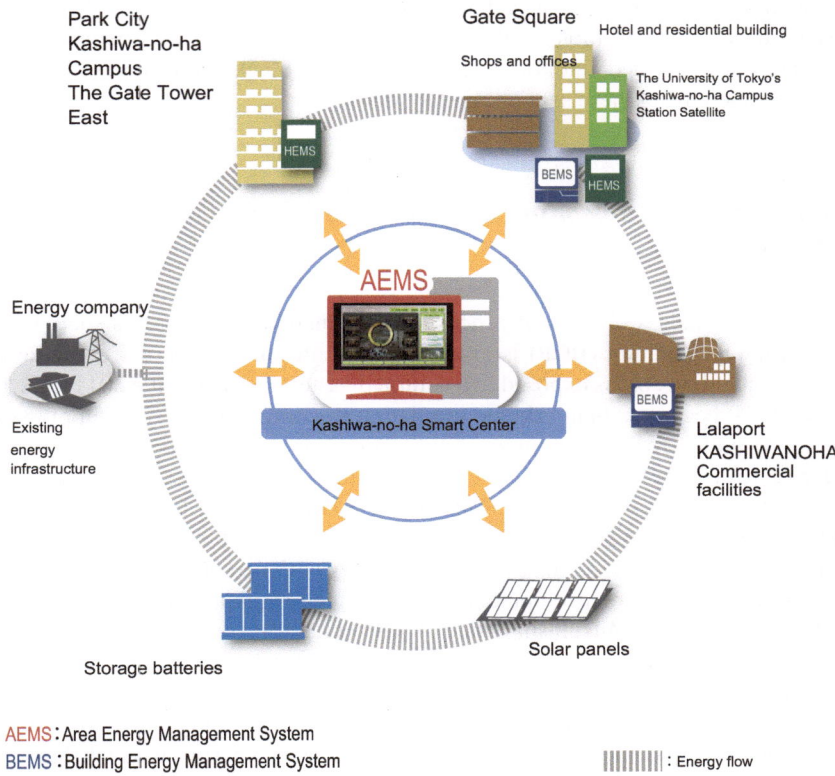

Fig. 3.2   Kashiwa-no-ha Smart City's Area Energy Management System (AEMS). Source: Mitsui Fudosan, *Kashiwa-no-ha Smart City* (Mitsui Fudosan 2019)

could provide energy interchange across an area spanning roads and railways, as well as roll out the system necessary to do so, without needing to apply for permission to provide power sharing under the Electricity Business Act.

Fujisawa Sustainable Smart Town is situated in what was once an industrial estate of approximately 19 hectares (planned population: 3000). The project is spearheaded by Panasonic. With a view to developing interdependent energy management, Panasonic equipped the area with 3 MW solar power systems and 3 MW storage batteries, and each (detached) house with a smart HEMS. Energy-saving technologies were also introduced. Each house stores energy using lithium storage batteries, generates energy using solar panels and energy farming, saves energy by using LED for all lighting, and uses water-efficient toilets and showers. Guidelines are distributed to ensure that the residents use these technologies effectively. Although area-wide energy management is the main focus, town management extends to a broad mix of services that support residents' quality of life. These services are designed to promote ecological sustainability and the well-being of residents. They are provided autonomously. The building blocks of a smart city in this case are formed using these services and systems.

The Smart City Concept in Business Continuity Planning for Urban Cores

An urban core houses clusters of offices and commercial facilities. As such, business continuity planning (BCP) is essential to ensure that businesses continue to operate if the urban core is exposed to a natural disaster. Following the 2011 Tohoku disaster, BCP became a major theme in relation to community energy supply systems, along with efforts to minimize carbon emissions and save energy.

One of Tokyo's key business districts is Chiyoda ward's Otemachi–Marunouchi–Yurakucho area (abbreviated as *dai-maru-yu* in Japanese and OMY in English). OMY has been undergoing redevelopment, and recently, the area has seen the development of swanky public spaces, including commercial facilities and customer attractions. This redevelopment is underpinned by an area management system driven forward by Mitsubishi Estate (The Council for Area Development 2019; OMY Area Management Association 2019). OMY has adopted the smart city concept in an effort to balance carbon reduction with BCP. Its smart city initiatives include energy saving in buildings, highly efficient community air-conditioning, fire-resistant architecture, and expanded rollouts of green energy (Inoue 2012).

Another example is Nihonbashi, a neighborhood in central Tokyo. Nihonbashi has a redevelopment zone, where redevelopment is led by Mitsui Fudosan. Mitsui Fudosan has introduced a cogeneration (combined heat and power) system as part of an effort to develop an energy infrastructure grounded in local power supply and heat generation. The building blocks of a smart city in this case are carbon-reducing physical infrastructure coupled with business continuity planning (Nakade 2017).

The Japanese Model of Smart Communities and Smart Cities

The above examples serve as models of the smart city in Japan. In the case of Yokohama, Kitakyushu, Keihanna Science City, and Toyota, there is a CEMS established in an existing urban environment, and there are locally anchored schemes for managing such a system. In the case of Kashiwa-no-ha Smart City and Fujisawa Sustainable Smart Town, there is a CEMS established as part of a large urban development project. In the case of Nihonbashi and OMY, there is a management system that extends to business continuity planning for the urban core's commercial cluster.

In each case, there is a CEMS tailored to the particular functions of the residential and commercial zones, under which innovative energy systems are introduced in accordance with the area's attributes and challenges. The above cases also suggest a smart city model that focuses on addressing local challenges; specifically, the above cases feature services designed to enhance residents' quality of life, and in the commercial areas, these services also include those designed to enhance safety and convenience.

On the one hand, we see energy management technologies (technologies that integrate IT with energy supply systems) being piloted, and we see the smart city model emerging with the practical implementation of these technologies. In non-energy sectors as well (such as in transport and healthcare), we see new services underpinned by IT-based management systems progressing from the test bed to the commercialization stage. That said, we have yet to identify any model to describe how these services and the management systems integrating different sectors can be commercially applied in multiple cities. As such, there is a need to reach out across different sectors, which requires us to go beyond the smart city framework.

3.4 Sustainable Cities and Smart Cities

Community Visions and Government-Led Projects

The smartification of urban/community energy management, including the Japanese cases described previously, has significantly propelled the evolution of smart cities. However, smart city initiatives in designated zones, such as new development zones, have limited ripple effects. Although there are no particular conditions on how large or populated a community or city should be to become a smart community or smart city, the selection of areas for smartification, as well as the scope of the system coverage (such as the CEMS), is arbitrary. As the selected areas become more energy efficient and their services improve, the task that arises is how to propagate the benefits of these technologies to areas beyond the existing smart cities. In an attempt to tackle this task, the government has focused on showcasing urban development models for others to imitate. This task requires communities to develop best practice models of urban (re)development. As it happens, many existing smart cities/communities took on this torchbearer role. Developing pioneering models for others to follow is arguably one of the social roles of smart community/city initiatives.

To position a smart community/city model project as part of a local community's overall strategy is essential for another reason: it helps clarify the community's general vision or master plan, identifies the roles to be played by each hard or soft initiative in this vision, and formulates strategies for implementing these initiatives. In this way, strategies can be implemented in a strategic and coordinated way, taking into account the ripple effects, including those related to addressing social issues such as global warming and aging population. This approach also ensures that the community uses government subsidies and grants as effectively as possible. The approach can be equated with government-led action in that government formulates a master plan for the community by clarifying its general approach and objectives, compiles an action plan outlining a specific strategy and concrete measures for accomplishing the master plan, and then works with residents and other local stakeholders in executing the plan.

Community Visions of Sustainable Development and Government Support

Government-led projects begin with a general vision for the community as a whole. When drafting this vision, it is crucial to understand the community's geographical and social attributes and incorporate them into the vision. It is also important to get a well-balanced big picture of the community and establish clear objectives related to sustainable development. The Sustainable Development Goals (SDGs), which the UN agreed to in September 2015, are relevant at the community as well as the national level, and these goals provide a universal framework that can be easily communicated globally.

The Japanese Government has indeed taken advantage of the SDG framework. In 2018, the Cabinet's Regional Revitalization Promotion Office solicited model projects under the "SDGs FutureCity" program and selected 29 municipalities. The office strengthened partnerships with a breadth of stakeholders, including communities already designated as an "eco-model city" or "FutureCity," and launched the "public–private platform for pursuing regional revitalization and SDGs" (Cabinet Office 2018).

Even before the UN established the SDGs in 2015 from 2000 onwards, the government had developed a program for supporting local initiatives; this was the program of designating communities as "eco-model cities" or "FutureCities." Under this program, 23 communities have been designated as "eco-model cities" and 11 as "FutureCities" (as of August 2018). These designers have established their own visions and strategies, and associated action plans (FutureCity Initiative 2019). Applications for FutureCity status began in 2011, the year of the Tohoku disaster. Of the 11 communities selected as FutureCities, 6 were in the disaster-hit areas. Some of the FutureCities have pursued smart community/city initiatives. For example, Kashiwa-no-ha (one of the smart cities discussed previously) used its FutureCity status to attract government funding toward the creation of an AEMS. Similarly, Higashi Matsushima used the funds to construct a "smart disaster-prepared eco town," which uses a CEMS to achieve energy self-sufficiency and enable energy supplies from neighboring communities in the event of an emergency.

Model Projects for Sustainable Urban Development

Ideally, such communities should take the initiative in analyzing their issues as well as their geographic and socioeconomic attributes. They should then address their issues, drawing on their particularities, underscoring their originality, and enhancing their appeal and vibrancy. In many cases, the communities should formulate a general vision and strategy and then advance concrete measures in partnership with private businesses, local organizations, and resident groups. On the other hand, when local governments take the lead in drawing up a vision, this vision may become overly dependent on bureaucratic processes. Accordingly, the Cabinet developed a platform to proactively support public–private partnerships and efforts to communicate the vision. Yet creating a vision alone is not sufficient; the key to its success lies in whether concrete measures are implemented effectively.

In the years ahead, it will be essential for public institutions, private businesses, and local residents to collaborate in implementing new initiatives. The government programs described above are, at least in part, intended to facilitate such public–private collaboration. Toyama City has implemented a series of measures intended to make the municipality a compact city. It introduced a light rail transit (LRT) system, which now forms the backbone of the city's infrastructure, and implemented road redevelopment projects based around the LRT. Additionally, the city has curtailed suburban expansion to make its urban environment more compact. Similarly, Kitakyushu City has implemented a smart community creation project in Higashida (Yahata-Higashi ward). Likewise, Shimokawa-cho (Hokkaido Prefecture) has advanced a renewable energy policy using its extensive forest resources as forest biomass. Earlier, we described such projects as "government-led." However, insofar as projects are supposed to demonstrate a framework for sustainable urban development, they should not just be government-led but rather be implemented collaboratively in public–private initiatives; they should also be conducted in a sustainable manner, so as to highlight the sustainable nature of the projects. The initiatives of the designated municipalities do indeed showcase successful models in each area; more importantly though, they serve as practical models describing how government, private businesses, and local residents can work together in implementing new transport or energy systems.

The Challenges of Japanese Smart Cities as Seen Through the Lens of Society 5.0

So far, we have identified two types of Japanese smart city initiatives: business-led initiatives conducted in conjunction with large-scale urban developments and government-led initiatives that are anchored within the vision statements of municipalities. How do these two types of smart city initiatives appear when viewed through the lens of Society 5.0? The key issue is whether these smart cities are, or can potentially become, compatible with the principle of the people-centric society. Can smart cities evolve to become capable of delivering goods and services that are highly customized to diversified and latent needs? To this end, smart cities require a new approach; rather than being led either by private businesses or by public institutions, they need to be led by citizens or based on citizen participation.

As mentioned in Sect. 3.3, the Japanese smart city model involves the use of a cutting-edge CEMS. The use of such a system as a means of practically implementing the smart city concept is certainly not without considerable value. However, it also exposes the fact that the Japanese smart city model is technology-led, being predicated upon the introduction of new technologies and systems. For smart cities to address the issues that urban populations commonly face, they need to be more citizen-friendly, use sensors and IoT-based technology, and be oriented more toward the vision of Society 5.0. On this score, we can look to examples of smart cities in the EU and the USA. These smart cities differ from the smart cities/communities in Japan, which are driven by the smartification of energy systems.

3.5   From Citizen-Led Smart City to Society 5.0

Smart Cities in the EU

The EU has supported the development and implementation of smart cities pursuant to the European Commission's medium-term vision "Europe 2020" (ratified in March 2010) and a program forming part of this vision, "Horizon 2020," which is the EU's largest program for supporting research and innovation, both financially and otherwise (Horizon 2020 runs from 2014 to 2020). In 2015, the European Commission launched the European Alliance for IoT Innovation (AIOTI) (Nomura 2017; NICT Europe Center 2017; Oshima 2016). Such institutional backing has yielded countless smart city models in Europe. These models feature a broad array of smartification—not only smart energy, but also smart transport, smart distribution, smart waste, and many other smart systems. In the following section, we focus on the example of Barcelona.

Smart City: Barcelona

Barcelona (population: 1.6 million) is the capital of Spain's Catalonia region. The city is renowned for its artistic heritage; it was the home to famous artists such as Pablo Picasso and Joan Miró, and it features many of Antoni Gaudí's buildings. Since hosting the 1992 Olympics, Barcelona has attracted attention for its subsequent economic growth and, more recently, has served as a model of a European smart city. The city uses sensors to monitor urban data. The data is relayed to citizens/users through apps or linked with community services such as transport systems and waste collection (see Fig. 3.3). This technology has enabled Barcelona to improve its traffic fumes and noise pollution, for which it was once notorious. Sensors installed around intersections monitor the air and noise pollution, and the readings are made freely accessible as open data. If readings in an intersection are high (indicating heavy pollution), the traffic signal patterns are adjusted so that vehicles flow through without stopping, thus lowering the traffic fumes around the intersection.

Unlike in Japan, many cities in the EU do not require people to obtain a parking certificate before owning a car. As these cities often lack adequate parking facilities, many parking spots are located by the sides of roads. Consequently, drivers who want to park on the side of the road must spend considerable time hunting for a space. To address this problem, Barcelona introduced a smart parking system. Asphalt-embedded sensors monitor whether the spaces are occupied, and drivers use apps to access this data and identify where the vacant spots are. These sensors are equipped with a battery and transmitter, and they emit signals indicating whether the space is vacant or occupied. These signals are overlaid on street maps in smartphone apps, allowing drivers to view the information in real time.

Other examples of smartification in Barcelona include smart lighting (streetlights that react to the presence of people), smart waste management (roadside waste containers use sensors that monitor when they are full), and smart cycling. An open-source platform called Sentilo connects the sensor data to the city's open data portal (Sentilo 2019). Sentilo has attracted attention for how it makes the data freely accessible globally. Barcelonan initiatives such as this have the potential to be adopted in other cities around the world.

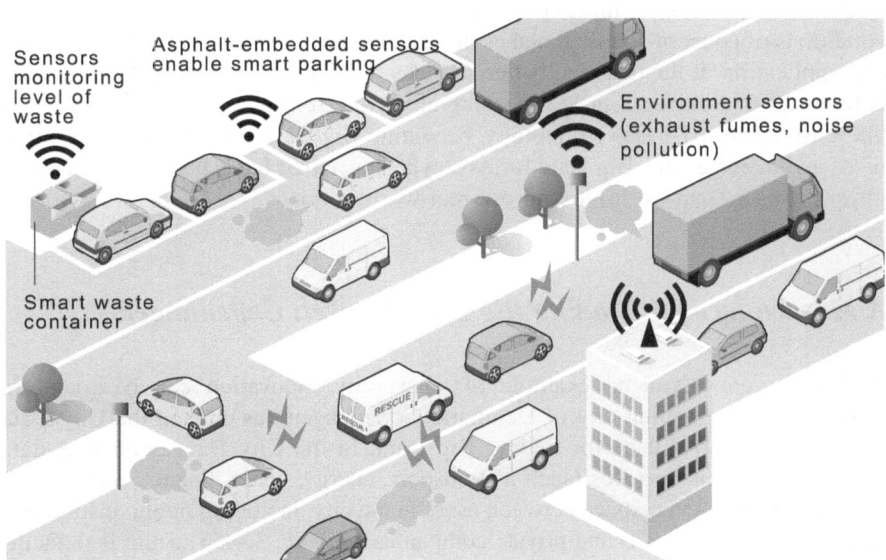

Fig. 3.3 Conceptual image of a sensing city

One intriguing Barcelonan initiative is Wallspot. Wallspot is an online tool that shows the locations of wall spaces available for legal graffiti. Barcelona had a problem with illegal graffiti. Some of the graffiti had artistic value, so the city released Wallspot to indicate legal graffiti spaces. Once artists have finished their work in these spaces, the painting is maintained for 1 week, after which it is removed and the wall space becomes available once again. Legal graffiti spaces are erected in public parks and their locations are advertised on Wallspot. The scheme has proven successful in reducing illegal graffiti in the city. Wallspot also helps graffiti artists connect with the local community; for example, it organizes graffiti events and displays works that have attracted interest among Barcelonans.

Transport is another area in which Barcelona has innovated. Tourists once complained about the city's confusing bus network. The city reorganized the network into vertical and horizontal lines, making the system much more intuitive. It also started displaying the waiting time for each bus service at bus stops. Additionally, Barcelona installed 500 free-charging stations for electric cars and scooters.

Smart City and Sensing City: Santander

Another Spanish city that has made an impact on the smart city scene is Santander (population of 180,000), the capital of the Cantabria region. Santander launched the "SmartSantander" project in 2010, earning the city EU funding. This funding was used to actively roll out sensor-based services that minimize personnel and service costs.

Santander is a key example of an EU city that uses a citizen-level approach to resolve local problems—more specifically, an approach that uses sensors to monitor conditions of concern and then makes the data freely accessible, allowing commercial application of the data and better services. The general thrust of this approach is to establish a citizen/user-led "sensing city." In a sensing city, data are gathered via sensors and IoT-based technology, becoming Big Data. The platform that organizes and manages this Big Data forms a cyberspace that feeds back the data to the physical space (real world) to improve real-world services.

A Marketplace for Trading Big Data Market: Copenhagen

An even more advanced example of smart city innovation can be found in Copenhagen, the capital city of Denmark. Copenhagen has created the City Data Exchange, a marketplace for trading Big Data. In the City Data Exchange, data related to different services (such as transport, energy, water, finance, and events) are exchanged in cyberspace between users in the city, including public institutions (such as the city council) and private companies (see Fig. 3.4). The aim is to facilitate the integrated use of the data, create new opportunities for businesses to trade in the data, and reduce the city's carbon footprint. The project emerged in the con-

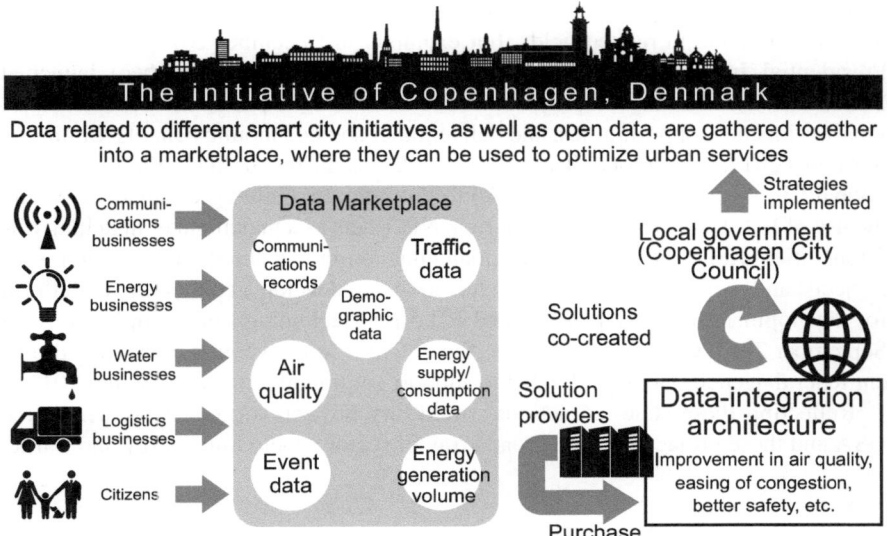

Fig. 3.4 Copenhagen's city data exchange

text of Copenhagen's policy objectives. Copenhagen set a goal of becoming the world's first carbon-neutral city by 2025. It then set a numerical target: to reduce its carbon emissions from the 2014 level of 2 million tons to 1.2 million tons.

To achieve this target, Denmark launched the Copenhagen Cleantech Cluster project (now known as CLEAN) to establish a cluster for introducing innovations in eco-technology. In 2014, CLEAN outlined a vision of digital infrastructure for gathering public–private data and analyzing Big Data in an ecologically effective way. In May 2016, the project launched a marketplace for trading data under a software-as-a-service model, allowing a wide range of organizations to purchase, sell, and share the data.

Although Northern European is a frontrunner in data sharing, even here, a data marketplace—businesses placing their data on the market and exchanging data with other companies—remains a new frontier. Businesses participating in consortiums have shown interest in a data marketplace, but they remain cautious about initiating trade in one. Hopefully, there will be more activity in the years ahead.

Smart Cities in the USA

In the USA, many of the smart city initiatives are driven by national policies and programs. Under the American Recovery and Reinvestment Act of 2009, the Obama Administration invested vast sums of federal money into the construction of a smart grid and energy-related digital technology, driving forward test beds and rollouts.

Since then, there has been a flurry of new initiatives to support R&D and infrastructural development in related fields. For example, in December 2013, White House Presidential Innovation Fellows (Geoff Mulligan and Sokwoo Rhee) launched SmartAmerica Challenge, a project that demonstrates the potential of IoT to create jobs and business opportunities, and deliver other socioeconomic benefits. Similarly, in August 2014, the National Institute of Standards and Technology (NIST) launched Global City Teams Challenge (GCTC) to promote the building of smart cities and the use of IoT. Under the GCTC program, NIST acts as a matchmaker matching different cities with common problems, matching common technological development projects, and matching cities with organizations, to develop a collaborative platform for developing smart city projects and IoT-based technology in multiple cities. In September 2015, Obama launched the Smart Cities Initiative, in which many different federal agencies are coordinated to support community efforts (Nomura 2017).

Reflecting these state-led actions, smart city projects are emerging across the USA and these projects cover a broad array of sectors, including energy and traffic.

Smart Cities in Maui, Hawaii

One example of a smart energy system is that of the Hawaiian island of Maui (population: approximately 150,000). From 2011 to 2016, energy stakeholders from Japan and Hawaii collaborated in a test bed project called JUMP Smart Maui ("the Japan–US Maui Project"). Hawaii was reliant on fossil fuel for around 90% of its energy, and it had set the goal of switching entirely to renewable energy sources by 2045. However, the state faced a challenge in relation to this task: because renewables fluctuate widely according to the weather, rolling them out on a large scale would destabilize the power grid. JUMP Smart Maui sought to demonstrate a method for stabilizing Maui's power grid in such a rollout. It integrated the island's wind power network with systems for charging and discharging all-electric vehicles in such a way that peak power use could be curtailed and vehicles could be charged during times of surplus energy. Eighty electric vehicles were used in the test bed, and electrical discharges from 14% to 31% of them yielded effective energy resources during peak times. Thus, the project managed to integrate electric vehicles as part of a flexibly dispersed power storage system. In doing so, it demonstrated that such a system is effective for stably managing renewable energy-sourced power in an enclosed locale such as a small island.

Sensing City: Chicago

Of all the American smart city models, the one that is most advanced in terms of open data is Chicago, Illinois (population: 2.7 million). In 2013, the city launched the Chicago Tech Plan. The Chicago Tech Plan consists of two foundational strategies: "next-generation infrastructure" and "every community a smart community." It also consists of three growth strategies: "efficient, effective, and open government," "civic innovation," and "technological sector growth" (Chicago Tech Plan 2019). As part of the "next-generation infrastructure," Chicago launched the Array of Things (AoT) initiative, in which the city installs sensors along the city streets to monitor real-time data on the urban environment and then makes the data freely accessible as open data. The data include temperature, humidity, barometric pressure, carbon monoxide level, ambient sound intensity, vibration, and pedestrian and vehicle traffic (Array of Things 2019). The sensors are mounted on light posts, and they house modules and other systems designed in a collaborative project, the leading members of which were the University of Chicago and the Argonne National Laboratory.

The data collected by the AoT sensors are open and freely accessible to businesses, researchers, citizens, and entrepreneurs. As of 2016, there were 42 sensors installed with plans to have 500 installed by 2018. To protect privacy and security, the data, together with the hardware and software, are regularly reviewed by an external, independent team (the Technical Security and Privacy Group). Because the Big Data yielded from the AoT enables real-time tracking of the urban environment, it empowers citizens to check conditions during disasters (such as floods) as well as the environmental conditions; it also has the potential to spawn new ideas for how to use the data. In this way, Chicago has cultivated a civic tech community, one in which Chicagoans take the initiative in leveraging cyberspace (in this case, sensor-based open data) in such a way as to benefit their physical space (real world).

Official Open Data Portal: San Francisco

San Francisco (population 0.8 million) is well known for its efforts to open up municipal data. In 2009, San Francisco launched an official open data portal called DataSF (DataSF 2019). DataSF contains a broad array of open data, including that related to urban planning, transport, housing, crime, and disasters. The city by the bay has also launched numerous apps for using this data, including an app that maps the city's buildings in 3D and an app related to real estate information.

However, open data poses challenges to the city government. The data must be constantly updated, and there must be an ongoing process for evaluating performance metrics, such as the time it takes to update the data. Additionally, while the data are free to use, the task of managing the data puts a constant strain on municipal budgets. The city government is introducing measures to address these problems so that it can keep the data open.

Some Japanese cities have taken a similar route. Fukuoka and Aizuwakamatsu, for instance, have launched official open data portals together with apps that allow citizens to access municipal data (Fukuoka City 2019; Aizuwakamatsu City 2019).

Challenges in Getting from the Citizen-Led Smart City to Society 5.0

The above case illustrates the general trajectory of the Western smart city scene: data on the city/community's issues are gathered at the level of citizens, and solutions to these issues are implemented by citizens or with the engagement of citizens. In other words, data related to the city/community's issues are opened to the public in cyberspace (the data are collected using sensors and opened to the public, or government data are made accessible on an official open data portal), and these data are then used to benefit the physical space (real world) by creating new services and business opportunities geared toward improving the environment. Thus, pioneering Western cities are already making fledgling attempts at bringing about the cyber-physical convergence to which Society 5.0 aspires. However, these attempts are still contained within cities, and they are limited to a particular cluster or limited to a particular sector or service.

As for the Japanese smart city scene, many smart cities/communities have emerged from test bed pilot projects pertaining to a particular system (such as the energy system) in a particular city neighborhood or city. Stated differently, Japanese smart city initiatives are limited to a particular area and a particular sector or service (see Fig. 3.5).

To progress from these smart city initiatives to the supersmart society of Society 5.0, where cyber and physical spaces converge, we must overcome several hurdles. First, the scope of the test bed projects must be enlarged to encompass entire cities and the entire society, and the projects must be liberalized. To this end, the regulatory climate must be eased, and the test bed process must be clarified and streamlined. Assistance from across government departments will be necessary, along with financial support, where necessary.

More work needs to be done also to engage citizens and users and to prepare a climate that continuously facilitates bottom-up, grassroots initiatives. Moreover, as the Western case studies testify, it is essential to form a platform for facilitating public–private–academic collaboration.

In addition, there must be innovative schemes to collect data on local issues coupled with support for business startup ideas that use such data. The key factor that will determine whether the scope of individual initiatives can be expanded to the local community as a whole is whether the city in question creates a mechanism that integrates new business ideas within the local industrial ecosystem.

Once we see Society 5.0 as the logical extension of smart city initiatives, the technical and institutional challenge should become clear: we need an information integration architecture that integrates data and information related to multiple services (such as transport, energy, and social welfare). In other words, the challenge is to build an architecture that links information from different fields. This challenge is discussed in the next chapter.

In overcoming this challenge, we should find two routes for advancing the smart city concept, which in Japan is indelibly associated with the technological clout of

	Japan	USA	EU
2007			■ Strategic Energy Technology Plan
2008	☐ Prooject for facilitating infrastructural measures in low-carbon-footprint urban development (MLIT, MOE) ☐ Ecological urban development project (MLIT) ☐ Eco-model cities (CO) Obihiro, Shimokawa, Iida, Yuzuhara, Minamata, etc.		
2009		■ American Recovery and Reinvestment Act of 2009) Dubuque 2.0 (Dubuque) DataSF (San Francisco)	■ Directive 2009/28/EC of the European Parliament and of the Council of April 23, 2009 on the promotion of the use of energy from renewable sources ■ EU climate and energy package Amsterdam Smart City (The Netherlands)
2010	☐ Next-generation energy and social systems testbed (METI) Yokohama, Toyota, Keihanna, Kitakyushu		☐ Europe 2020 Smart Santander (Spain)
2011	☐ FutureCities (CO) Kashiwa-No-Ha, Shinchi, Higashi Matsushima, Toyama, etc. ☐ Smart community vision proliferation (METI)	JUMP Smart Maui (Hawaii)	■ Energy Efficiency Plan 2011 ○ EU Smart Cities Information System Smart City Lyon (France)
2012	■ Eco-City Act (Low Carbon City Act) ☐ Project to promote urban development, residential, and transport models that create, store, and save energy (MLIT) ☐ Project to promote ICT-based urban development (MIAC)	■ Digital Government Strategy	○ European Innovation Partnership on Smart Cities and Communities Copenhagen Connecting (Denmark)
2013	☐ Project to promote models of resident-led carbon reduction planning (MOE) ○ Council to promote ICT-based urban development (MIAC) Smart City Aizuwakamatsu	☐ Smart America Challenge ○ Smart Cities Council Chicago Tech Plan	
2014		☐ Grobal City Teams Challenge ■ Digital Accountability and Transparency Act	☐ Horizon 2020 ■ Digital Agenda for Europe 2020 Copenhagen Cleantech Cluster (Denmark)
2015	☐ Project to promote ICT-based urban, human, and employment development (MIAC)	☐ Smart Cities Initiative ☐ Smart City Challenge	○ Alliance for IoT Innovation (AIOTI) Paris intelligente et durable (France) Smart City Berlin (Germany)
2016	■ 5th Science and Technology Basic Plan ■ Comprehensive Strategy on Science, Technology, and Innovation ■ Basic Act on the Advancement of Public and Private Sector Data Utilization	Smart Cincy (Cincinnati)	
2017	☐ Project to promote data-based smart cities (MIAC) Sapporo, Takamatsu, Kakogawa, Urawamisono, etc.	Smart Columbus	■ General Data Protection Regulation (GDPR)
2018	☐ SDGs FutureCity (CO) Kanagawa Prefecture, Kamakura, Maniwa, Iki, Oguni, etc.		

| General legend | Smart city case | ■ ...Legislation, state plan, etc.
☐ ...Project
○ ...Other | Abbreviations for Japanese public institution | METI = Ministry of Economy, Trade and Industry
MIAC = Ministry of Internal Affairs and Communicatinos
MLIT = Ministry of Land, Infrastructure, Transport and Tourism
MOE = Ministry of the Environment   CO = Cabinet Office |

Fig. 3.5 Synopsis of smart city trends in Japan, the USA, and the EU

private businesses. First, there will be more business- and government-led progress. Second, we will see more citizen-led or citizen-involved progress. Simply put, overcoming the challenge will help pave these two tracks. Once the groundwork is laid, citizen groups can start gathering, analyzing, and applying urban data (such as sensor data). In other words, we will see a society where Big Data analytical tools are deployed to make life in the city more comfortable and convenient as well as to empower local communities to solve their issues. Such an outcome would signify that the smart cities of today are making progress in cultivating the society to which Society 5.0 aspires. In the not too distant future, we should see such activity in communities across the land.

References

Aizuwakamatsu City (2019) Ōpundeita no torikumi (Open data initiative). https://www.city.aizu-wakamatsu.fukushima.jp/docs/2009122400048/. Accessed 4 June 2019

Aram A (2017a) Global innovation report: microgrid market in the USA. Hitachi Rev 66(5):454–458. https://www.hitachi.com/rev/archive/2017/r2017_05/Global/index.html. Accessed 4 June 2019

Aram A (2017b) Global innovation report: Beikoku ni okeru maikuro guriddo (Microgrid market in the USA). Hitachi Hyoron (Hitachi Rev) 99(2):166–171. http://www.hitachihyoron.com/jp/archive/2010s/2017/02/02Global/index.html. Accessed 4 June 2019. The third paragraph of the quote is translated from the Japanese version of Hitachi's comments; the prior paragraph is taken directly from the English text of the same comments

Architectural Institute of Japan (ed) (2014) Sumātoshiti jidai no sasutenaburu toshi/kenchiku dezain (Sustainable cities and architectural designs in the smart city era), Shokokusha

Array of Things (2019). https://arrayofthings.github.io/. Accessed 4 June 2019

Cabinet Office (Regional Revitalization Promotion Office) (2018) Chihō sōsei SDGs kanmin-renkei purattofōmu ni tsuite (Public-private platform for pursuing regional revitalization and SDGs), June 2018. https://www.kantei.go.jp/jp/singi/tiiki/kankyo/pdf/sdgs_pura_gaiyo.pdf. Accessed 4 June 2019

Chicago Tech Plan (2019). https://techplan.cityofchicago.org/executive-summary/. Accessed 4 June 2019

DataSF (2019). https://datasf.org/opendata/. Accessed 4 June 2019

Fujisawa Sustainable Smart Town Association (2019) Fujisawa SST (Fujisawa sustainable smart town). https://fujisawasst.com/. Accessed 4 June 2019

Fukuoka City (2019). Biggudeita ōpundeita no katsuyō suishin ni muketa torikumi (Initiative to promote the use of Big Data and open data), http://www.city.fukuoka.lg.jp/soki/joho/shisei/BDODkatsuyou.html. Accessed 4 June 2019

FutureCity Initiative (2019). http://future-city.jp/en/. Accessed 4 June 2019

Ikeda S, O'oka R (2014) Nihonkokunai ni okeru sumātoshiti sumātokomyuniti jisshō jigyō no saishin dōkō (Recent trends in testbed projects for smart cities/communities in Japan). J Inst Ind Sci 66(1):69–77

Inoue S (2012) Daimaruyū (ōtemachi-marunouchi-yūrakucho) chi'iku ga egaku sumātoshiti to wa (What is the smart city vision for the Otemachi-Marunouchi-Yurakucho area?). Institute for Global Environmental Strategies, workshop on urbanization knowledge platform for low-carbon cities, July 26, 2012. https://www.iges.or.jp/jp/archive/gc/activity20120726.html. Accessed 4 June 2019

Japanese Business Federation (Keidanren) (2017) Revitalizing Japan by realizing Society 5.0: action plan for creating the society of the future, February 14, 2017. http://www.keidanren. or.jp/en/policy/2017/010_overview.pdf. Accessed 4 June 2019

Ministry of Economy, Trade and Industry (2019) Jisedai enerugī/shakai shisutemu kyōgikai ni tsuite (On the council for next-generation energy and social systems). http://www.meti.go.jp/ committee/summary/0004633/. Accessed 4 June 2019

Mitsui Fudosan (2019) Kashiwanoha sumātoshiti (Kashiwa-no-ha smart city). https://www.kashi-wanoha-smartcity.com/. Accessed 4 June 2019

Nakade H (2017) Nihonbashi sumātoshiti: Enerugī no jiritsuka to chisanchishō ni yoru saigai ni tsuyoku kankyō ni yasashii machizukuri (Nihonbashi's smart city: liberalizing energy and establishing local production local consumption as a basis for disaster-resilient and eco-friendly urban development), Institute for Building Environment and Energy Conservation, 37–5(218):8–11

National Institute of Information and Communications Technology (NICT) Europe Center (2017) Ōshū ni okeru IoT to sumātoshiti no kenkyūkaihatsu ni kansuru dōkō (European trends in IoT and smart city R&D). https://www.nict.go.jp/global/4otfsk000000osbq-att/a1489129184837. pdf. Accessed 4 June 2019

Nikkei BP Clean Tech Institute et al (2011) Sekai sumātoshiti sōran 2012 (Overview of the world's smart cities, 2012), Nikkei Business Publications

Nomura A (2017) Yūzā doribun inobeishon ni yoru sumāto na machizukuri ni mukete: Kaigai ni okeru 'sumātoshiti 2.0' e no torikumi (Toward smart urban development based on user-driven innovation: smart city 2.0 initiatives overseas), Japan Research Institute. JRI Rev 8(47):101–139. https://www.jri.co.jp/MediaLibrary/file/report/jrireview/pdf/9939.pdf. Accessed 4 June 2019

OMY Area Management Association (2019). http://www.ligare.jp/. Accessed 4 June 2019

Oshima K (2016) Ōshū no sumātoshiti to biggudeita (Smart cities and big data in Europe). J Archit Build Sci 131:1690

Sentilo (2019). https://ajuntament.barcelona.cat/digital/en/digital-transformation/urban-technol-ogy/sentilo. Accessed 4 June 2019. http://www.sentilo.io/. Accessed 4 June 2019

The Council for Area Development and Management of Otemachi, Marunouchi, and Yurakucho (2019). http://www.otemachi-marunouchi-yurakucho.jp/. Accessed 4 June 2019

Tsuchiya Y (2015) Sumātoshiti no keisei yōken to jitsugen hōsaku ni kansuru kenkyū (The elements of and policies for effectuating a smart city), Ph.D. Thesis, Tokyo Metropolitan University

Yamamura S (2015) Sumātoshiti wa dō tsukuru? (How do you make a smart city?), NSRI sensho (Nikken Sekkei Research Institute Anthology), Kousakusha

Chapter 4
Integrating Urban Data with Urban Services

Ryosuke Shibasaki, Satoru Hori, Shunji Kawamura, and Shigeyuki Tani

Abstract This chapter provides an overview of the architecture for integrating urban information. It describes how urban information should be integrated and how this integration can result in collective optimization of services. The integration of spatial and temporal information represents the initial approach for the architecture of social and technical integration. A much better metric to improve is user satisfaction, wherein users are the individuals and businesses in cities.

Further, three key channels for integrating information are discussed here. The first channel is the set of interfaces that enable systems to operate symbiotically. Specifically, this indicates the designing interfaces that allow businesses and services to communicate and interact with each other such that all of them operate not only independently but also as part of a larger organic system. The second channel is the set of social systems that recalibrate the rights and responsibilities concerning the use, management, and protection of data. Technology that can enable organizations to use personal information without compromising on data privacy and data principles is introduced. The last channel is a measure of the quality of life (QoL).

The original version of this chapter was revised: This book was inadvertently published with the incorrect license type CC BY 4.0 and the Open Access License has been amended throughout the book to the correct license type CC-BY-NC-ND. The correction to this chapter is available at https://doi.org/10.1007/978-981-15-2989-4_9

R. Shibasaki (✉)
Center for Spatial Information Science, The University of Tokyo, Tokyo, Japan
e-mail: shiba@csis.u-tokyo.ac.jp

S. Hori
Social Systems Engineering Research Department, Center for Technology Innovation—Systems Engineering, Research & Development Group, Hitachi, Ltd., Tokyo, Japan
e-mail: satoru.hori.fy@hitachi.com

S. Kawamura
Security Research Department, Center for Technology Innovation—Systems Engineering, Research & Development Group, Hitachi, Ltd., Tokyo, Japan
e-mail: shunji.kawamura.wz@hitachi.com

S. Tani
Social Systems Engineering Research Department, System Innovation Center, Research & Development Group, Hitachi, Ltd., Tokyo, Japan
e-mail: shigeyuki.tani.dn@hitachi.com

This has already been discussed in the second section of Chap. 2. In this chapter, we discuss a theoretical framework for measuring the QoL using human sensing.

Keywords Anonymous analysis technology · City OS · Data platform · Personal data protection · Symbiotic autonomous decentralized system

4.1 Architecture for Integrating Urban Information

Two Approaches to Integrating Urban Information

Cities are large population centers with a defined space, and they host a cluster of shared and coordinated activities from which goods and services are efficiently produced and consumed simultaneously. Underlying this dense cluster of activity are infrastructure services, including those related to travel, distribution, communications, energy supply, waste management, and water supply and treatment. To ensure that these services operate effectively, cities must devote considerable resources to gathering data and quantitatively managing it. For years, water, electric power, and gas supplies; road and rail transport; and many other services have been run by control systems that are based on real-time data. Data gathering and analysis, as well as use of the findings to control services or guide decision-making, represent the core activities of urban information management.

Data/information related to each infrastructure service—water, power, gas, road and rail transport, and the like—are managed separately. However, if we could spatially map all of these infrastructure services together, we could understand how they interconnect or, in some cases, conflict. For example, water supply, sewage systems, power lines, gas pipelines, road networks, and subway networks often overlap with each other (above and below the ground). Workers who lay wastewater pipelines must understand the locations of the power lines and gas pipelines. The team must also try to minimize disruption to road traffic. When conducting public works, if the team can collate data on the locations of each infrastructural facility and related public works and adjust worksites and schedules accordingly, it will be possible to prevent accidents and minimize inconvenience to road users.

A system to facilitate such coordination has in fact been in operation for some 20 years now. In this system, parties share information on the whereabouts of facilities occupying road sections and information on planned construction/maintenance works so as to consolidate and visually map out the information. This system is underpinned by common regulatory stipulations, which provide that the relevant parties must use such a system to coordinate their operations. The system constitutes the basis for the parties to input, manage, and share data. The regulatory stipulations also provide that the costs of the system must be shared among the relevant actors. As this example illustrates, to ensure that parties share data and information, there must be an information-sharing system in place, and there must also be rules governing how the system works as well as a model for sharing expenses.

IT is used to optimize urban activities and infrastructure. Often, however, these optimizations have been carried out separately without considering the interconnections between them. We have mentioned how important it is to consider the different infrastructural facilities occupying road sections, but this also holds true for different transport services. For example, to ensure that passengers can transfer smoothly from a train to a bus, or to ensure that the mobility-as-a-service trend (MaaS: the integration of different transport services to allow users to seamlessly transfer between them) gains traction, it is essential to integrate information pertaining to each of the relevant services. But how should the information be integrated? The idea of integrating this information might seem like an obscure idea at first, but it becomes clearer when we consider how much urban services overlap spatially. At transport nodes, such as train stations, rail, bus, and taxi, services intersect. If the data related to each of these transport services are mapped together spatially and temporally, this will enable the visualization of the interconnections and conflicts between them. Consequently, we will get a clearer picture of how we should integrate them and how this integration can optimize the services collectively. Such spatial and temporal information integration represents the first of two essential approaches for social and technical integration architecture (see Fig. 4.1).

The second approach concerns the question: What should be optimized? If we are just focusing on an individual system, the optimization targets should be relatively easy to define. They would include things like travel time and operating costs. However, the situation gets more complicated when we consider the whole set of systems. Perhaps the obvious thing to optimize in this case would be the combined operation cost, but this metric does not much help us to envisage future urban systems. After all, an effective way to optimize costs is to lower standards in many cases. A much better metric to improve is user satisfaction—users being the individuals and businesses in cities. This raises the question: What is user satisfaction?

Users invariably seek services that offer comfort and joy, that take away the strains of life, that meet their demands down to the smallest detail, and that are deliverable on demand. Such services are not as fanciful as some might think; indeed, they are already being provided in some respects. Uber, for instance, has demonstrated how mobility services might work in a society where automated driving has proliferated. Uber vehicles are not automated as such, but the principle is similar: vehicles are dispatched upon the user's request such that the user does not need to drive. Likewise, if you stay at a luxurious hotel, you can get a very real taste of a comfortable lifestyle afforded by complete automation. At this hotel, you will not need to do any housework, and you can get something to eat or drink at the click of a finger. These high-level services could be delivered at a dramatically reduced cost through information integration or with AI and Internet of Things (IoT)-based technologies. Thus, these conveniences are technically feasible and people desire them. Yet, should we be seeking to extend such an ideal environment to everyone (as our optimization mission) simply because it is feasible and desired?

Rosemarie Parse proposed a theory of nursing called "Human Becoming" (Parse 1998). She developed this concept (which she had originally named "man-living health") after questioning the extent to which absolute nursing care and

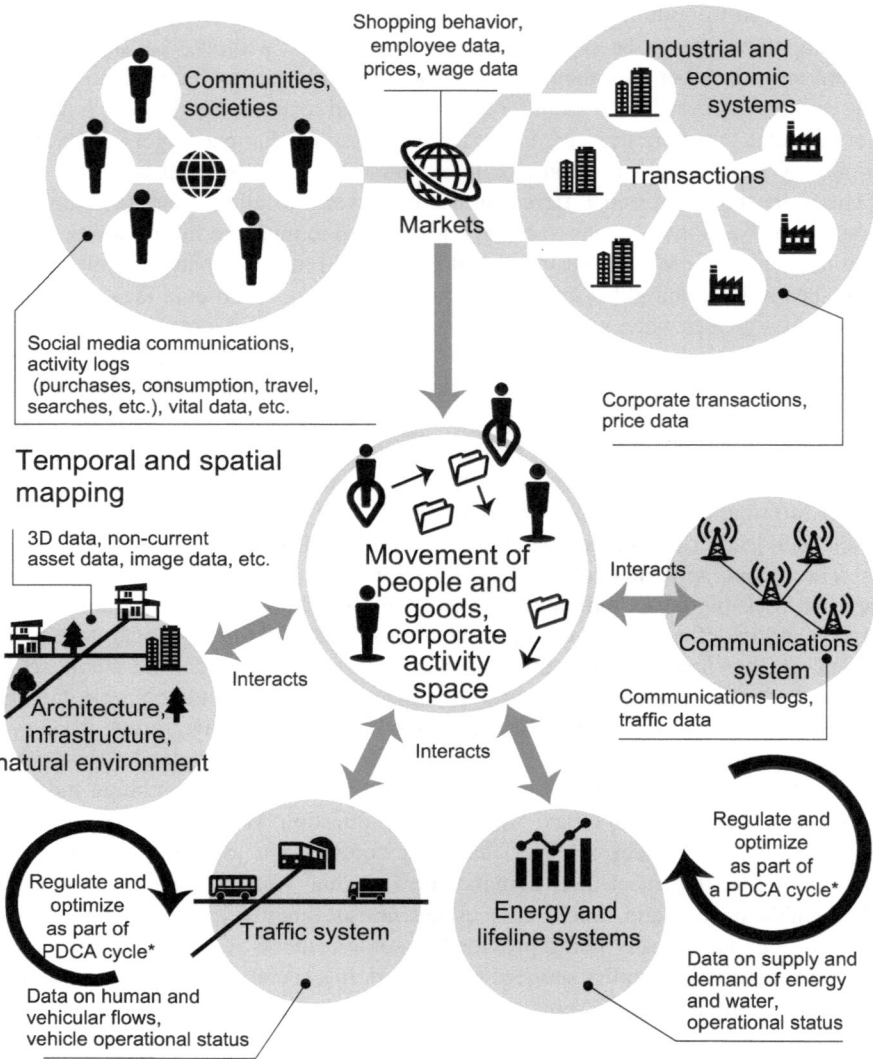

Fig. 4.1 Data landscape in urban management: temporal and spatial sharing enables interactivity

life-preserving medical care are necessary, and how much it is in the patient's interest. Human Becoming assumes that humans derive meaning from freely choosing their actions and experiences, upon which they grow and change (or *become*). The theory also posits that humans make these free choices based on the information they gather in their intersubjective interactions with the environment.

Ideas such as these should be considered in a process of broad public discussion and test bed programs. As our society accumulates a body of discourse and experience, as well as a body of digital data, we can gain increasing insight into what we should be optimizing, and then develop the environment accordingly.

Channels for Integrating Information

There are three key channels to integrating information. The first channel is technical; it concerns the interfaces that enable systems to operate symbiotically. Specifically, this means designing interfaces that allow businesses and services to communicate and interact with each other such that they all operate independently but also as part of an organic whole. Section 4.2 introduces an approach to achieve such symbiosis: "Symbiotic autonomous decentralized system," a concept that focuses on integrating existing services.

The second channel is institutional: it concerns social systems that recalibrate the rights and responsibilities concerning the use, management, and protection of data gathered from different systems, and that consider the outcomes, costs, and impacts of the data use. Currently, the notion that applies to property in general—that the one who acquires the property enjoys all the rights connected with it—applies to information and data too. However, issues concerning data privacy challenge this notion. If you own or manage someone's personal information, you will potentially affect this person in some way. You cannot use the data against the wishes of the data principal. Ultimately, therefore, the data principal must have a right to be involved in the process of using or managing his or her data. A prominent example of this principle is the right to data portability (the right to transfer one's personal data from one data controller to another), which is enshrined in the EU General Data Protection Regulation (GDPR). The principle also applies to the example we discussed earlier of assets occupying road sections. That is, to ensure the appropriate use of publicly owned roads, data must be shared to the extent necessary, and the different parties must compromise their own interests in pursuit of the wider collective interest. There must be an ongoing discussion on how to rectify the excessive emphasis on sectional interests and how to achieve the wider interests of the whole. Such a discussion will help enshrine data rights and stakeholder responsibilities as a cornerstone of governance. Once we have such a discussion and common understanding, we must stipulate the rights and responsibilities that should apply when different services are integrated. We must also design a model for how the costs will be shared. These institutional steps are essential, for without them, the systems, as technically feasible as they may be, will fail to gain traction in society.

The bulk of the urban data will be the personal information of users (including city residents, workers, and visitors). Societies around the world have now established the principle that personal information can be entrusted to reliable organizations and individuals, which will use the data to identify ways to improve and optimize urban services. This principle has been put into practice with schemes such as personal information banks and personal data stores, but these schemes have met with difficulties. The reason they struggle is related to the irrecoverability of losses from personal information accidents, such as when the data are leaked. If we have technology that can enable organizations to use personal information without compromising data privacy, data principals will be much more inclined to give their consent to the use of their data, which will open up many new possibilities for using data. Section 4.3 introduces such a technology, one that encrypts personal data so that they can be analyzed while retaining the anonymity of the principal.

The third channel concerns governance structures under which society can continually fine-tune the separate systems, the architecture integrating these systems, and the methods for managing data. To this end, it is necessary to refer to quantifiable measures of citizens' QoL (the ultimate metric of society-wide optimization) and the distribution thereof. QoL is, by definition, a measure of quality of life. We already discussed this in the section on Habitat Innovation framework (Sect. 2.2). In Sect. 4.4, we will discuss a theoretical framework for measuring QoL: human sensing.

Ongoing Issues

Once the objective function is determined, approaches to system integration and optimization will start producing results in many different settings. However, a number of issues must first be addressed. For example, how can cities establish objective functions when they have so many different actors whose interests may come into conflict? How do they form consensus among these different interests? How should private businesses and public institutes (including local authorities and residential communities) share data and information as part of the process of forming consensus and determining common goals? These issues have only just started to be debated. That said, a number of ideas and schemes have already come forth. For example, data portability (the ability to gather, manage, and access one's own data) has gained worldwide traction. Likewise, there is a groundswell of support for the idea that public institutions can use data collected by private organizations in ways that can benefit communities. In the years ahead, it will be crucial to implement concrete measures by which citizens, communities, businesses, and local governments can amass a body of experience. It will also be important to increase opportunities for these actors to autonomously discuss specific strategies for addressing the issues.

4.2   Symbiosis of Urban Systems: Symbiotic Autonomous Decentralized System

A Vision of Service Cooperation

How can we make different services cooperate? In practice, it is unfeasible to develop a single megalithic system that binds all the services together. After all, different services operate under their own separate systems; therefore, to consolidate them all into a single system would take far too much money and time. Another problem is that every time you change a service, you would need to recalibrate the system as a whole, which will again cost time and money. Rather than developing a megalithic central system, it would be much more feasible to make existing systems cooperate.

How then do we accomplish this feat? To make the systems cooperate, they must be modified or a new function must be added. However, it may not be realistic to implement all these changes in one go. Instead, the systems should cooperate in stages. Once the systems are cooperated, each system will need to be constantly updated. During updates, the overall systems must remain unencumbered, even if the system in question is temporarily suspended. Given these requirements, it becomes clear that we should not couple the systems together so tightly as to make them all dependent on one another. Instead, we should aim for loosely coupled systems, wherein the composite systems operate independently but interact with each other so as to benefit the systems as a whole.

Autonomous Decentralized System

A concept that can help in the design of such a system is "autonomous decentralization." This term is based on an observation about living organisms. An organism is made up of numerous cells. These cells are subsystems that operate independently, but they also interact with each other to enable the system as a whole—the organism—to function. Likewise, an autonomous decentralized system features a set of subsystems, each of which operates independently but in a way that contributes to the system as a whole (Hitachi 2019). This is the very principle we must follow in designing integration architecture.

Let us consider this autonomous decentralized system in a little more detail. For subsystems to operate interactively and harmoniously, they must communicate with each other. When a subsystem passes a message, it will usually pass the message to a particular subsystem. Likewise, when you send someone a package, you attach a label onto the package indicating the person's address. The situation becomes more complicated, however, when a new subsystem is added. Under the conventional approach, if a message needs to be passed to the new subsystem, the sender subsystem must be modified. Likewise, you would need to update your

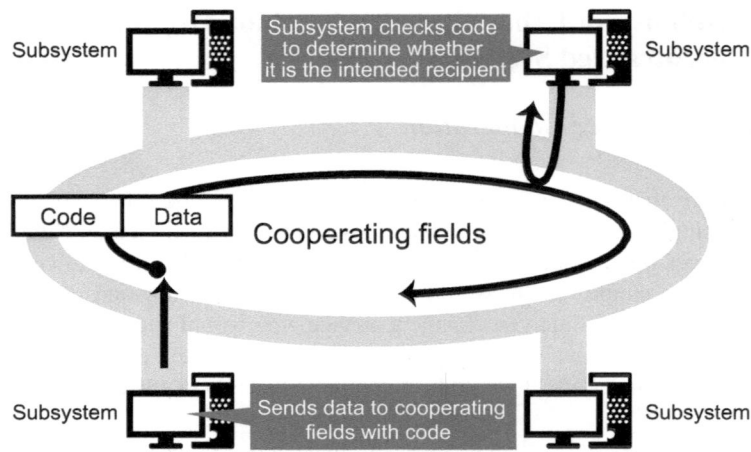

Fig. 4.2 Autonomous decentralized system

address book without delay if the domicile of one of the entries changes. However, an autonomous decentralized system avoids the need to modify existing subsystems each time a new subsystem is added, thanks to a mechanism described in the next paragraph. Such a system requires no address book.

Figure 4.2 illustrates how the subsystems pass messages to each other. Suppose that a system is made up of ten subsystems. When Subsystem A needs to pass a message to Subsystem B, it deposits the message in a designated collection area (without indicating Subsystem B as the intended recipient). The subsystems (other than Subsystem A) then check the collection area to see whether the message is intended for them. The message will only be picked up by the subsystem that determines that it is the intended recipient. How does Subsystem B verify that it is the intended recipient of the data? The data that each subsystem processes are structured in a way that accords with that subsystem's particular function. Thus, each subsystem checks the structure of the data, and if the structure corresponds to the subsystem's function, the subsystem will conclude that it is the intended recipient of the data.

We turn now to considering forward compatibility. What happens when Subsystem B gets an upgrade, becoming "Subsystem B+"? Subsystem B+ processes the same kind of data as Subsystem B. As such, Subsystem B+ will determine that it is the intended recipient of the data that Subsystem A sent. Therefore, there is no need to modify Subsystem A (Subsystem A does not need to be told to change the recipient from Subsystem B to Subsystem B+). Thus, an advantage of the autonomous decentralized model is that the subsystem can be easily modified in stages, making large-scale system development much simpler.

The places where the subsystems deposit their messages are called "cooperating fields." The subsystems are all linked to the cooperating fields, so they can send all deposit data in and collect data from the cooperating fields. We mentioned that subsystems determine whether the data are for them by checking the structure of the data. The process is actually even simpler in practice. There is a coding system to describe different types of data content, and the messages deposited in the cooperating fields are each labeled with a code. Subsystems need only to check the code to determine whether the data are intended for them.

Symbiosis of Systems

The principles of autonomous decentralization we have just outlined are also applicable to the cooperation of the systems for different sectors or services. If a cooperated system is akin to an organism, then a cooperated group of systems is like a symbiotic community of independent organisms. Hence, Hitachi has coined the term "symbiotic autonomous decentralized systems" (Irie et al. 2016).

To illustrate this principle, consider the autonomous decentralized cooperation of transport management systems, including those related to trains, buses, and taxis. Each of these systems operates independently, and in the course of their operations, the systems post real-time data on their transport operations onto cooperating fields. Then, another system can view the data posted onto the cooperating fields, and use this data to shape an efficient transport plan for the entire group of systems. This plan will then be posted onto the cooperating fields. This process enables transport services to be much more efficient than a situation where each transport system formulates plans separately.

This symbiosis can occur across different sectors to enable even more efficient service delivery. To take transport and energy services as an example, as part of an effort to promote electric vehicles, transport plans can be arranged so that electric vehicles recharge during hours when energy prices are lower. Such cross-sector cooperation will help optimize society as a whole, bringing us closer to the supersmart society.

4.3 Personal Data Protection: Anonymous Analysis Technology

Personal Data Leaks

It is from Big Data, such as data on people's shopping history and sensor data, that we gather and analyze data, and extract new information. One method for exploring Big Data is association rule learning, which can help identify relations between variables. For example, an analysis of shopping histories might tell us that when people buy diapers, they tend to buy beer as well. Such relationships are called "association rules." The discovery of a good association rule can prove very useful in guiding product planning, promotional campaigns, and store layouts. Big Data analytics are conducted not only in marketing but also in a wide range of fields. For example, sensor data are analyzed to determine the causal factors of accidents.

In the years ahead, Big Data will increasingly be used to shape services that rely on personal data. Data analysts will be increasingly entrusted with data from massive samples, including data on the TV programs people watch every day, the websites they visit, their shopping history, and the shops and restaurants they have visited. They will use the data to find association rules that indicate trends, such as people's tastes and preferences, the products that groups with similar tastes and preferences tend to purchase, and where they go to purchase these products.

How, though, can we entrust our personal data to these organizations with peace of mind if there are no solid security measures in place? The best method currently available is one that involves blacking out key information, such as the principal's name and address, so as to prevent the principal from being identified. In the years ahead, we will see more use of encryption as a means of enhancing personal data security. If data are sent for analysis in an encrypted form, it helps prevent the risk of the data being exploited for malicious purposes. Encryption is already used in the Internet; when we submit personal data online, for example, the data are encrypted. Anonymous analysis technology (Naganuma et al. 2014) enables analysis to be conducted without the encrypted data being decrypted and returned to its original state. In this way, it significantly offsets the risk of unauthorized access or malicious leaks. Naturally, the way to decrypt the data is concealed from the party analyzing the data.

Anonymous Analysis

Generally speaking, if you entrust a data analyst with your personal data but provide the data in an encrypted form, the analyst will be unable to process the data, as only you know how to decrypt it. If on the other hand you use anonymous analysis technology, the data analyst could find association rules in the data even in its encrypted state. From these association rules, the analyst could identify certain products to recommend to you, but these product recommendations would expose tastes and preferences that you may be uncomfortable sharing. No one wants to broadcast all their tastes and habits to the world. The notable thing about anonymous analysis technology is that the product recommendations themselves are encrypted so that the analyst cannot understand them. You alone can decrypt the recommendations using your initial password. Such technology is now in the process of development.

Figure 4.3 illustrates the technology. First, you encrypt your personal data with a password and upload the data to a cloud server (a remote data center). When you send an encrypted query, the data center collates the query with your encrypted personal data and issues an encrypted response. Because the data and query remain encrypted on the cloud server, no one other than you can understand them. Finally, you receive the encrypted response and decrypt it using the same password and check the content.

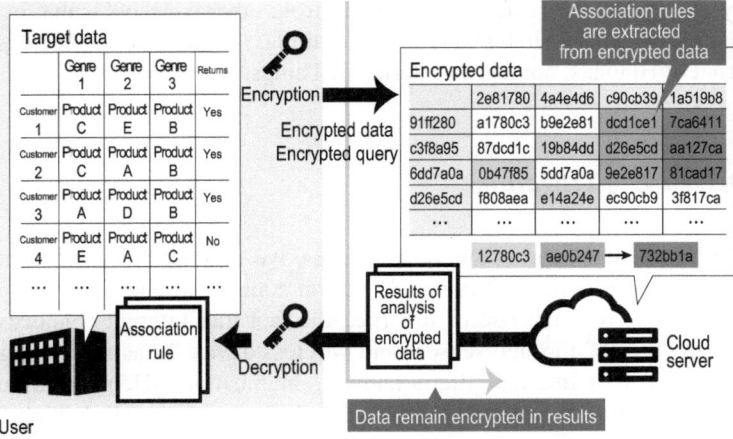

Fig. 4.3 Anonymous analysis service

Anonymous Analysis Using Searchable Encryption

The key technology in anonymous analysis is searchable encryption. Searchable encryption allows encrypted word searches to be performed on encrypted documents such that neither the words searched for nor the documents analyzed will become known. In conventional searches, one must temporarily decrypt the data to find the target terms. With searchable encryption however, the number of times search terms appear in a document can be identified while keeping the data encrypted. Searchable encryption can therefore be used to statistically analyze data and find association rules without breaching anonymity.

The ability to access data without compromising privacy will significantly expand the use of data, because it will become much easier to obtain the consent of the principals toward the use of their data. Thus, anonymous analysis—technology that enables data to be analyzed securely (in that the data remain encrypted)—has an important contribution to make in protecting privacy.

4.4 Measuring Happiness: From the Internet of Things to the Internet of Humans

In the manufacturing sector, the Industrial Internet, Industrie 4.0, and similar concepts have been proposed as ways to connect machinery, robots, and other things to the Internet, to go digital, and to realize significant leaps in productivity. Japan's vision of Society 5.0, on the other hand, proposes a human-centered society, one that delivers comfort and happiness through a high degree of cyber-physical convergence.

The key factor that differentiates Society 5.0 from other concepts is that in Society 5.0, connectivity extends to humans as well as things; in other words, Society 5.0 has an Internet of Humans, not just an Internet of Things.

IoT-Driven Digitalization

With the rise of the Internet and smartphones, we now live in what some dub a "ubiquitous network society," in which we can connect whenever, wherever, with whomever. Moreover, the rise of IoT (i.e., the fact that Internet connectivity has extended to "things" through sensors and wireless devices) has further digitalized society (Ministry of Internal Affairs and Communication 2015). As discussed in Chap. 1, a vision of Industrie 4.0 was outlined as part of the High-Tech Strategy 2020 Action Plan for Germany (Industrie 4.0 Working Group 2013). Industrie 4.0 proposed a vision of digitized supply chains that use data collected from IoT-based devices to innovate manufacturing processes.

In 2012, General Electric Company (GE) outlined its vision of the Industrial IoT or the Industrial Internet. The idea of the Industrial Internet is to connect manufacturing hardware with advanced analytical software so as to dramatically reduce cost and generate new value (Japan Business Federation 2016). In each of these ideas, a core role is played by cyber-physical systems, which characterize the fourth industrial revolution. Such systems are already being used to fully digitize manufacturing via IoT, and significant gains in productivity are starting to be realized.

Society 5.0's Novel Concept: Human Centrism

Whereas economic systems have traditionally derived their competitiveness from their ability to consolidate "things" and money in ways that improve efficiency, in Society 5.0, the source of economic value will instead lie in people and data according to the Growth Strategy 2017 (Growth Strategy Council 2017); moreover, data related to people's wisdom and behavior will generate value and create a society in which individuals from all walks of life can actively contribute. The way in which we use data gathered from the Internet of Humans, in addition to that gathered from the Internet of Things (which was the main player in Industrie 4.0 and the Industrial Internet), will be of critical importance in our efforts to usher in Society 5.0. The next section explores the benefits of the IoH, as well as the problems it entails.

What Is the Internet of Humans? It Starts with Human Sensors

Why do we speak of humans, as well as things, being connected in a network? Let us consider some examples. Hitachi recently conducted research into wearable sensors that measure happiness.

According to Yano et al. (2015), happiness correlates significantly with performance. Happy people are 37% more productive and 300% more creative than unhappy people. They also have more friends and longer healthy life spans. There are economic effects too: companies with happy workforces report higher earnings per share (Yano et al. 2015).

In a Hitachi study, the activity patterns of call center employees were measured using wearable sensors (as shown in Fig. 4.4). The data, which represented a total of over 1 million days, indicated that the sensors tended to emit signals suggesting happy moods during times of active behavior patterns (Yano et al. 2012). A call center team with high happiness levels won 34% more business than a team with lower happiness levels.

These findings imply that even in these times of automation, our mental well-being remains key to economic success. They further suggest that we will gain new economic growth opportunities if we use the IoH to digitize human activity and create a people-centric society that supports happiness.

The Benefits of IoH

We have just described one aspect of IoH—the use of wearable sensors to monitor human activity. From the perspective of the people wearing these sensors, this is a passive process; their sensor data are gathered and used by others. How should IoH

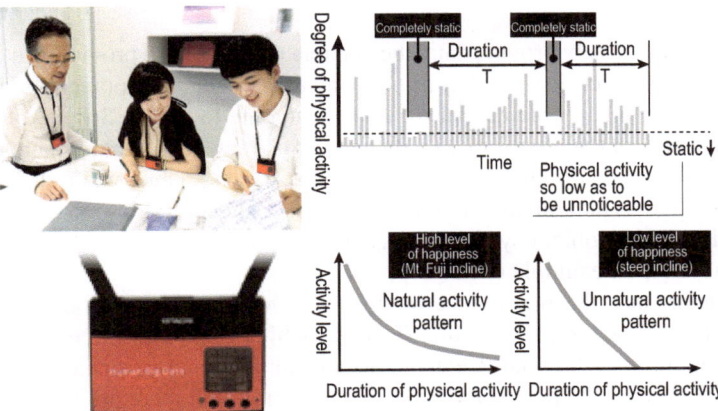

Fig. 4.4 Physical activity patterns associated with happiness (measured by wearable sensors)

operate in Society 5.0's economic system? Habitat Innovation avers that active citizen participation is key to such an economic system, and data are no exception. The ideal form of IoH according to Habitat Innovation is one where citizens provide data actively, not passively.

To get an idea of active data provision, we can consider energy supply and consumption. To control energy supply, energy companies monitor the amount of energy we consume each day and use the data to forecast future energy demand. The energy companies must maintain a constant balance between demand and supply to ensure a continuous supply of energy at rated voltage and to avoid outages. To this end, they must forecast demand accurately and flexibly adjust supply to accommodate sudden demand changes. What would happen if citizens actively provide energy consumption forecasts for the next day, the next week, or the next month? The uncertainty over future demand would be minimized, allowing energy companies to forecast demand and adjust supply at much lower cost, thus freeing up economic potential.

An example of a pilot project that accorded with this approach to IoH is Kutsuplus ("call plus"), an on-demand minibus service launched by Helsinki City Transport (HSL) (Toyota 2015). Kutsuplus was a service that matched passengers' travel needs with a minibus driver. Users would enter their starting point, destination, and desired arrival time into a smartphone app. The app would then match the query with a suitable boarding point, disembarkation point, and timetable. Unlike in conventional bus services, minibus drivers would adjust their routes and timetables to reflect the data that users provided. Minibus drivers could then deduce the optimum routes to meet user needs and dynamically customize the service accordingly. The result was an economical transport tailored to citizens' demand (see Fig. 4.5).

The Problems That IoH Entails: Hurdles That Must Be Overcome on the Way to Society 5.0

When citizens actively provide data, they enable urban service providers to dynamically adjust their services, which in turn unleashes economic potential. However, the IoH also entails problems that we must overcome to realize Society 5.0. These problems concern the reliability of data and the privacy of the data principals.

The problem regarding data reliability is that datasets can contain dubious data that unscrupulous individuals submit with malicious intent. E-commerce and review sites address this problem by allowing users to rate the reliability of the content. The "like" button is an example of this.

As for privacy, the importance of this issue has been underscored by the EU General Data Protection Regulation (GDPR), which came into force in May 2018. The jurisdiction of the GDPR extends to all multinational companies around the world that process the personal data of EU residents, and there are tough penalties for multinationals who fall foul of the GDPR (EU GDPR.ORG 2019). Insofar as

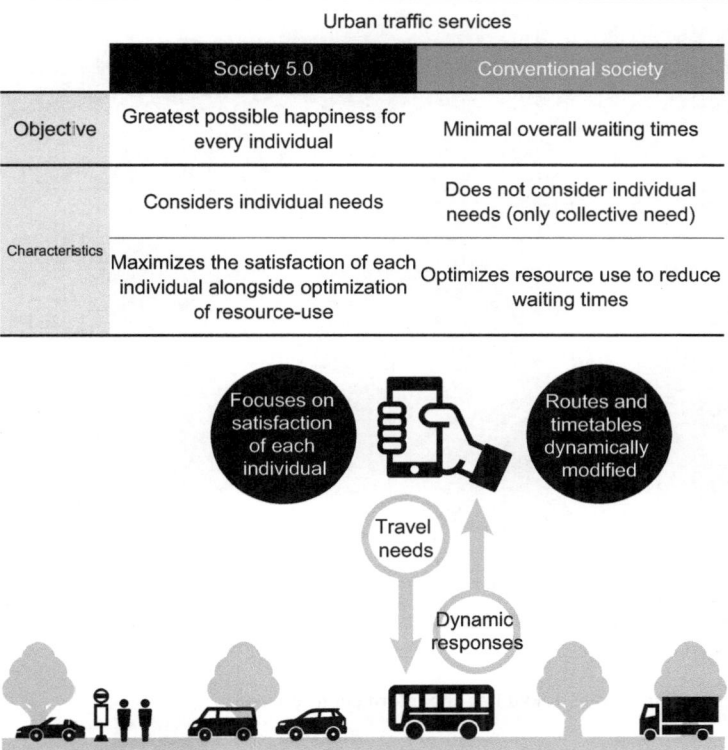

Urban traffic services		
	Society 5.0	**Conventional society**
Objective	Greatest possible happiness for every individual	Minimal overall waiting times
Characteristics	Considers individual needs	Does not consider individual needs (only collective need)
	Maximizes the satisfaction of each individual alongside optimization of resource-use	Optimizes resource use to reduce waiting times

Fig. 4.5 Urban traffic services in Society 5.0's people-centric society

Society 5.0 will be based on human connectivity, we must address the challenge of data privacy. We previously mentioned anonymous analysis technology; in the discussion below, we introduce an intriguing data privacy service.

In 2016, the major railway company Tokyu Corporation launched Station Vision (the registered trademark is "駅視 (Ekishi)-vision"), a service that relays station crowd-level data to users as image data. The data are based on video camera footage. If the service sent this footage to users unedited, it would violate the privacy of the individuals that appear in it. To avoid this problem, Station Vision replaces the images of the people with icons. Using Hitachi's people-flow analytics, Matsukuma et al. (2017) developed a system that detects people's walking direction and then replaces the images of these people with icons depicting their movement, as shown in Fig. 4.6. If the same icon was used for each person, it would be hard to intuitively understand the crowd dynamics, as the direction of each individual would be unclear. The use of direction-specific icons makes it much easier to gage crowd levels, and does so without compromising privacy. Station Vision thus provides a way for users to check the crowds at stations from their homes. When trains are delayed, such as during heavy snowfall, users can check the crowd

(Level of crowding)

Fig. 4.6 "Station Vision," Tokyu line app information service

levels at the stations and then decide whether to take an alternative route or to wait at home. The service also helps railway companies ease congestion at stations (Matsukuma et al. 2017).

As we have learned, IoH—human connectivity that enables the digitization of behavior and demand—has an important role to play in society. When citizens provide data actively, rather than passively, urban services (including energy and transport services) will be able to dynamically customize their services to users. This change will bring economic benefits, but to gain these benefits we must also address the privacy issue. In doing so, we will accelerate progress toward Society 5.0.

References

EU GDPR.ORG (2019) GDPR key changes. https://eugdpr.org/the-regulation/. Accessed 4 June 2019

Growth Strategy Council (Headquarters for Japan's Economic Revitalization) (2017) Growth strategy 2017: reforms aimed at achieving society 5.0, June 2017. https://www.kantei.go.jp/jp/singi/keizaisaisei/pdf/miraitousi2017_summary.pdf. Accessed 4 June 2019

Hitachi Webpage on the Autonomous Decentralized System (2019). https://www.hitachi.com/products/it/control_sys/platform/ads_net/index.html. Accessed 4 June 2019

Industrie 4.0 Working Group (2013) Recommendations for implementing the strategic initiative INDUSTRIE 4.0: final report of the Industrie 4.0 Working Group, April 2013. https://www.din. de/blob/76902/e8cac883f42bf28536e7e8165993f1fd/recommendations-for-implementing-industry-4-0-data.pdf. Accessed 4 June 2019

Irie N, Ohashi A, Onodera T, Kato H (2016) Information and control systems—open innovation achieved through symbiotic autonomous decentralization—. Hitachi Rev 65(5):13–19

Japan Business Federation (2016) Society 5.0 to IoT nado e no torikumi (Initiatives related to Society 5.0 and the IoT). https://www.jpo.go.jp/resources/shingikai/sangyo-kouzou/shousai/tokkyo_shoi/document/16-shiryou/03.pdf. Accessed 4 June 2019

Matsukuma N, Osawa T, Nukaga N, Otsuka R, Kato M (2017) Using people flow technologies with public transport. Hitachi Rev 66(2):145–149

Ministry of Internal Affairs and Communications (2015), 2015 White paper on information and communications in Japan. http://www.soumu.go.jp/johotsusintokei/whitepaper/eng/WP2015/chapter-5.pdf#page=13. Accessed 4 June 2019

Naganuma K, Yoshino M, Sato H, Sato Y (2014) Privacy-preserving analysis technique for secure, cloud-based big data analytics. Hitachi Rev 63(9):577–583

Parse RR (1998) The human becoming school of thought: a perspective for nurses and other health professionals. SAGE Publications, Thousand Oaks, CA

Toyota (2015) On-demand public transportation changing human mobility in cities. https://open-road-project.com/en/innovationreview/post_611/. Accessed 4 June 2019

Yano K, Lyubomirsky S, Chancellor J (2012) Sensing happiness: can technology make you happy? IEEE Spectr 49(12):32–37

Yano K, Akitomi T, Ara K, Watanabe J, Tsuji S, Sato N, Hayakawa M, Moriwaki N (2015) Measuring happiness using wearable technology—technology for boosting productivity in knowledge work and service businesses. Hitachi Rev 64(8):517–524

Chapter 5
Solving Social Issues Through Industry–Academia Collaboration

Atsushi Deguchi, Yasunori Akashi, Eiji Hato, Junichiro Ohkata, Taku Nakano, and Shin'ichi Warisawa

Abstract This chapter illuminates how Society 5.0 will transform our cities and lives through introducing the research works developed by industry–academia collaboration "H-UTokyo Lab," which is a joint undertaking by Hitachi and the University of Tokyo. In this chapter, researchers from the field of engineering discuss the basic thought process behind the research projects aimed at addressing social problems in each section, including those related to the aging population, the need to go carbon free, and the need to regenerate rural communities. In addition to the discussion, the researchers also describe the updates in technical revolution to

The original version of this chapter was revised: This book was inadvertently published with the incorrect license type CC BY 4.0 and the Open Access License has been amended throughout the book to the correct license type CC-BY-NC-ND. The correction to this chapter is available at https://doi.org/10.1007/978-981-15-2989-4_9

A. Deguchi (✉)
Department of Socio-Cultural Environmental Studies, Graduate School of Frontier Sciences, The University of Tokyo, Tokyo, Japan
e-mail: deguchi@edu.k.u-tokyo.ac.jp

Y. Akashi
Department of Architecture, Graduate School of Engineering, The University of Tokyo, Tokyo, Japan
e-mail: akashi@arch.t.u-tokyo.ac.jp

E. Hato
Department of Civil Engineering, Graduate School of Engineering, The University of Tokyo, Tokyo, Japan
e-mail: hato@bin.t.u-tokyo.ac.jp

J. Ohkata
Institute of Gerontology, The University of Tokyo, Tokyo, Japan
e-mail: okata@up.t.u-tokyo.ac.jp

T. Nakano
Department of Housing and Urban Planning, Building Research Institute, Ibaraki, Japan
e-mail: nakano@kenken.go.jp

S. Warisawa
Department of Human and Engineered Environment Studies, Graduate School of Frontier Sciences, The University of Tokyo, Tokyo, Japan
e-mail: warisawa@edu.k.u-tokyo.ac.jp

solve the social problems. Each section concludes with an illustration of the image of our lives in future in Society 5.0.

In Sect. 5.1 we provide an overview of social problems in Japan and then propose the basic approach to solve these. In Sects. 5.2–5.4, we explore the approach and direction for technology development that facilitate innovating cities and living spaces in relation to each of the following three propositions. Section 5.2 suggests ideas for better housing support for the 100-year life and development of technologies that are close to human and data-driven services. Section 5.3 introduces technologies that coordinate energy management at different tiers, that is, individual, building, and district level, to contribute to a carbon-free society, which enables people to use minimal energy without sacrificing their QoL. Section 5.4 proposes a data-driven urban planning method, which supports the local community to develop their own community improvement projects.

Keywords Energy and life management system · Data-driven urban planning · Habitat innovation · Healthy aging · Zero-carbon society

5.1 How Will Society 5.0 Transform Cities?

Society 5.0 is a people-centric society that resolves both economic and social issues while ensuring that people live comfortable and fulfilling lives. To that end, how must urban environments change? How should we try to change them? What must we do to effectuate such changes?

This chapter concerns a novel model developed by H-UTokyo Lab., which is a joint undertaking by Hitachi and the University of Tokyo. Rather than following the conventional model of industry–academia partnership, in which a university lab conducts joint research with a private firm, H-UTokyo Lab. seeks to solve social issues through industry–academia collaboration, which involves organizational integration between the firm and the university. Under this approach, researchers from both Hitachi and the University of Tokyo form working groups on separate themes, and work on technologies and policy proposals under these themes. In this chapter, we focus on the discourse related to these endeavors.

The First Thing to Change Is Values

In Chap. 2, we outlined the approach of Habitat Innovation. Habitat Innovation seeks to innovate cities and habitats without being beholden to prevailing social conventions. It is only through such a bold approach that we can transform society. To ensure that flexible, outside-the-box ideas gain traction, we must first and foremost replace, create, and revive values. More specifically, we must do the following:

1. Replace the prevailing values that have held us back
2. Create new values to release us from conventional frameworks by drawing on accumulated knowledge
3. Revive abandoned values

By doing so, we move closer to making Society 5.0 a reality.

It is only when we challenge and replace the very values to which we traditionally adhere that we will ignite the innovation necessary to create a people-centric society, which then will become a motivating force promoting Society 5.0. Stated differently, Society 5.0 is not the logical extension of today's society; Society 5.0 is a revolutionary break with prevailing ideas and practices.

Here, we will consider three propositions (together with solutions) concerning the urban habitat innovations that are necessary to bring about a change in mind-set and acclimatize society to new ways of thinking. The first is that elderly people should be enabled to continue living their own homes. The second is that people should have more choices in their living and working environments. The third is that local communities should take the initiative in identifying their attractive features. These propositions might seem obvious and straightforward, but it is very hard to implement them under conventional ways of thinking. They must be implemented however, if we are to build a people-centric society.

How can we use technological and social innovation—the kind that arises from the convergence of cyberspace and the physical space (real world)—to fulfill these propositions? What technology and policy changes are required for such ends? We will discuss technological approaches and policy proposals later (from Sect. 5.2 onward); first, we will clarify the meaning and challenges of each proposition from residents' perspective.

Enabling Elderly People to Continue Living Their Own Homes

As we mentioned in Chap. 2, Japan's population is graying at an unprecedented rate. There are growing numbers of elderly people living alone, and many elderly people themselves care for other elderly people, which has become a major social issue. The rate of aging is particularly high among the many suburban areas in the Greater Tokyo Area, as well as those in other metropolises in Japan, the development of which peaked during the high economic growth period. Many of the inhabitants of these neighborhoods have resided there since the neighborhoods first developed. With few new residents moving in, the resident population is either stable or declining, and there are an increasing number of vacant properties. Consequently, local services (such as those related to shopping and healthcare) are struggling to meet elderly residents' daily care needs. Hence, it will be no easy task to ensure that elderly people can continue to live in this housing, to which they are so accustomed.

Japanese life spans have lengthened to such an extent that the government has set out the "age of the 100-year life" as a national policy. Accordingly, to ensure that people can continue living comfortably in their familiar neighborhoods, there needs to be systems of support that reflect each resident's health conditions and the

circumstances of their neighborhoods. There must also be habitats that enable
elderly people to live independently, and so we must develop systems and technologies to achieve that end.

First, we must change the mind-set. Figure 5.1 breaks down conventional elderly
housing policy into residence status (single living—living with spouse/family members) and the convenience of the neighborhood (transport services, daily living
facilities, etc.). The X-axis indicates the former while the Y-axis indicates the latter.
Conventional policies in this area either encourage residents to relocate to more

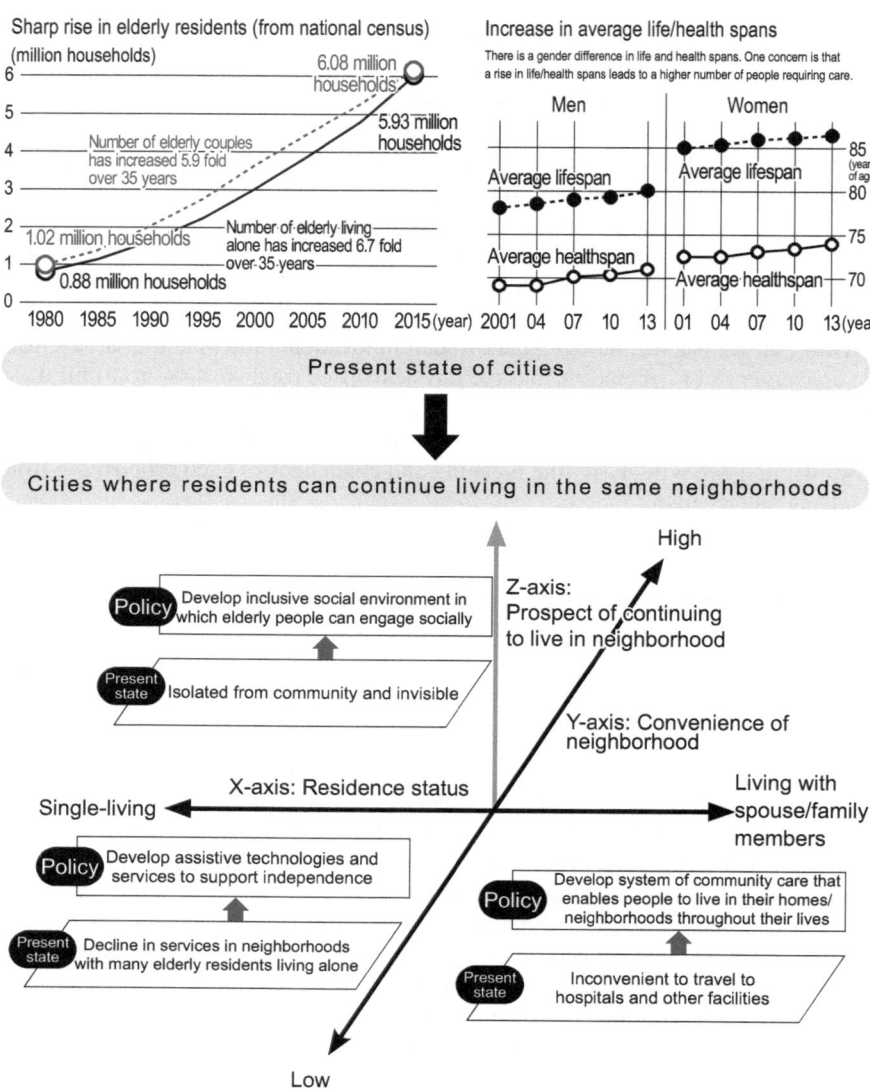

Fig. 5.1 Paradigm shift in supporting elderly people's housing

convenient local areas (the compact city model) or promote elderly care, including social welfare-based care and family-provided care. The fundamental issue though is how to create age-friendly living environments, where the elderly can continue living with peace of mind. Hence, we added a third axis, the Z-axis, to indicate the value of elderly residents continuing to live in the house in question. Using this metric, we must reassess conventional approaches to elderly housing, which are designed to reflect age-based changes, and start thinking about new solutions.

Insofar as Society 5.0's people-centric society is one in which people can continue living in the same neighborhoods, it must include cyberspace architecture that can help elderly people to live with peace of mind. Of importance here is the creation of an environment that preempts or minimizes dependency on care—an environment where elderly people's health is carefully managed on a daily basis so as to offset the risks of sudden injury, disease onset, and other risks to health and safety. It is also essential to develop assistive technology to create an environment customized to the person's lifestyle and thus encourage independent living. In Sect. 5.2 of this chapter, we suggest ideas for how this paradigm shift in values can lead to better housing support for the 100-year life and development of technologies that are close to human and data-driven services.

More Choice in Where You Live and Work

The traditional image of a metropolis is a place that houses a cluster of employment centers (such as offices and commercial facilities) in its center and residential areas out in the suburbs, or commuter towns (or "bed towns" as they are called in Japan). As property and land prices in the city center are high, many white-collar workers live in the less expensive suburban areas and endure a lengthy daily commute into the center. Figure 5.2 shows the distribution of the residential population and the employee population in the Greater Tokyo Area. The residential population distribution extends across the suburbs. The worker population, however, is concentrated in the center, reflecting the fact that the white-collar workplaces in the Greater Tokyo Area are clustered in central Tokyo. Thus, the majority live far away from their jobs; they work in central Tokyo, and live in the suburbs.

Figure 5.2 includes a graph with four quadrants in which the X-axis indicates where workers in the Greater Tokyo Area live (suburbs vs. central Tokyo) and the Y-axis indicates where they work (suburbs vs. central Tokyo). As of 2015, the Greater Tokyo Area has a working population (population of full-time or part-time workers) of 16 million,[1] with 14.5 million for whom the residential and work loca-

---

[1] "Greater Tokyo Area" refers to the municipalities within the "urbanized areas" and "suburban areas" as defined in the Greater Tokyo Area Development Plan. "Central Tokyo" refers to the 23 central municipalities (often called the "special wards"), Kawasaki City, and the bay area of Yokohama City. "Suburbs" refers to the municipalities in the "suburban areas." In both cases, the statistics are based on 2015 national census data concerning individuals aged 15 or older, who are employed or in school.

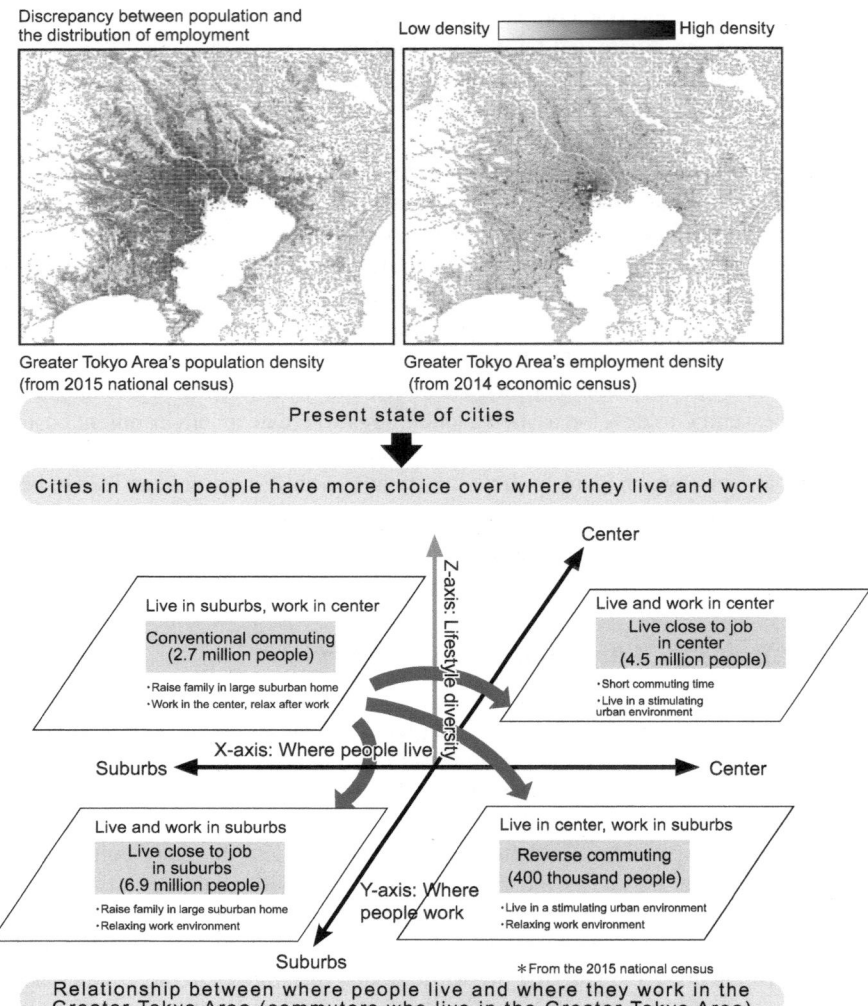

Fig. 5.2 Paradigm shift wherein people live and work in a metropolis

tions are known. Of these 14.5 million workers, an estimated 2.7 million live in the suburbs and work in central Tokyo. Only 400 000 do the reverse (live in central Tokyo and work in the suburbs), a ratio of 1:7.

This trend is reflected in train congestion patterns. During the morning rush, inbound trains (trains running from the suburbs into central Tokyo) are packed while outbound trains have few passengers. The reverse is true in the evening rush hour. Thus, the train networks in Japan's metropolises are still not used to full capacity. The Greater Tokyo Area has an extensive rail network compared to other metropolises around the world, and suburban rail lines radiate out from the terminus stations on the Yamanote line (the railway loop line in central Tokyo). Given this pattern, if more people were to live in central Tokyo and work in the suburbs, it would lead to more effective use of the rail network and raise the capacity of the Greater Tokyo Area.

There is some literature on "reverse commuting" (residents who commute to the suburbs) in overseas cities, including Paris and New York (Aguilera et al. 2009; New York Times 2008). Reportedly, some Google employees live in central San Francisco and commute to suburban Silicon Valley. Some suburban employment centers have also emerged in the Greater Tokyo Area, such as Futako-Tamagawa and Kashi-no-ha Campus. For those who work in engineering and development, it would make a lot of sense to work in a verdant and stress-free suburban environment as opposed to one of the high-rise towers clustered in the city center. Reverse commuting is, of course, contingent upon having affordable housing and land prices in the city center, but if people can work in cyberspace, that would remove the need for employment to be concentrated in the city center and would create greater scope for new lifestyles, such as living in the city center amid a wealth of leisure facilities (such as restaurants and cinemas) and commuting to the suburbs for work. Shifting from the traditional urban model of concentration and specialization to a model that emphasizes decentralization and diversification will be a vital step toward building a people-centric society that accommodates a variety of lifestyle options.

The key to realizing this shift in urban design is to disincentivize urban centralization. Traditionally, urban centralization is held to benefit society, in that having offices concentrated into the same geographical area offers greater economic efficiency and more efficient energy use. The downside, however, is that workers spend many hours a day commuting under packed conditions, depriving them of disposable time. This situation runs counter to the people-centric society of Society 5.0. There are a number of proposals for addressing the problems created by living far away from one's job. Telecommuting and small office/home office (SOHO) are examples of this. Thus, progress is being made in developing technologies and environments that remove the need to commute to a physical office every day; for example, workers can work in decentralized office environments connected in cyberspace. However, decentralized office environments lead to increased and less efficient energy consumption, problems that may deter progress. Thus, no fundamental breakthrough in this issue has yet been made. Recently, we have started seeing the practical application of energy management systems for Building/Community Energy Management Systems (BEMS/CEMS) , but to create environments that function flexibly and operate across the whole of the Greater Tokyo Area, it is necessary to develop technology based on energy management systems that is not tied to the geographical concentration of business clusters and, as such, can minimize energy consumption across the whole of society. In other words, we must apply the symbiotic autonomous decentralized system throughout the Greater Tokyo Area (more on this in Chap. 4).

The third section of this chapter introduces technologies that coordinate energy management at different tiers (individual, building, district, Greater Tokyo Area) to contribute to a carbon-free Japan. These technologies will help effectuate the shift from traditional urban models that emphasize economic efficiency toward an urban model that supports diverse lifestyle choices (and thus supports Japan's work style reforms) and enables people to use minimal energy without sacrificing their QoL.

Local Communities Taking the Initiative in Identifying Their Attractive Features

A staggering number of local governments in rural areas are at a loss in how to deal with shrinking populations. Many of these local governments have set out policies designed to maintain or increase the resident population and nonresident population (tourists, visitors, sojourners) by deploying the area's tourist resources and the abundant natural environment. However, given the long-term decline in Japan's overall population, local governments are fighting over an ever-smaller pie. A more fundamental task is illustrated in Fig. 5.3: local communities must examine past data on nonresident and resident populations and then derive a sustainable future vision that suits the locality. They must then implement effective policies to achieve the vision.

Some local communities have seen their resident population decline but have also seen their nonresident population increase. What should such communities be doing in their effort to make themselves sustainable? Should they seek to bolster the nonresident population by drawing further upon their attractive features? The Society 5.0 solution is to use cyberspace to analyze data (including plenty of case data) and compare different scenarios for alternative future visions. Cyberspace provides local communities a tool for making informed decisions on their future direction.

At present, however, local governments do not always have access to the simulation tools and data they require for conducting fact-finding surveys or forecasting the outcomes of policies. Local governments can ascertain the facts on the community concerned by accessing public statistics such as national and economic censuses. They could also access the basic city planning surveys prescribed under the City Planning Act or nationally conducted household travel surveys ("person-trip" surveys). The problem with these public resources, however, is that the surveys are carried out infrequently, at 5- or 10-year intervals. The granularity of the data is limited too; the data are collected in meshes or at a municipality, district, or subdistrict level. In other words, the public data do not adequately capture the dynamics of municipality, where changes occur frequently and locally. This data cannot, for example, help local governments forecast the outcomes of widening a road or laying down a new road, or running a precise after-evaluation of policies or pilot experiments. Local governments lack the data and analytical tools with which to smoothly run plan–do–check–act cycles. We no longer live in an age where local governments should wait upon the national government to supply them survey data; local governments should waste no time in accessing Big Data, including satellite imagery data and mobile spatial data, and then use this data to build cyberspace architecture.

To proceed with town building projects (such as road development) flexibly while forecasting the outcomes of the plans, local governments in the course of such projects should gather and use data on the project zones and the peripheral zones. To this end, local governments require platforms that integrate effective Big Data.

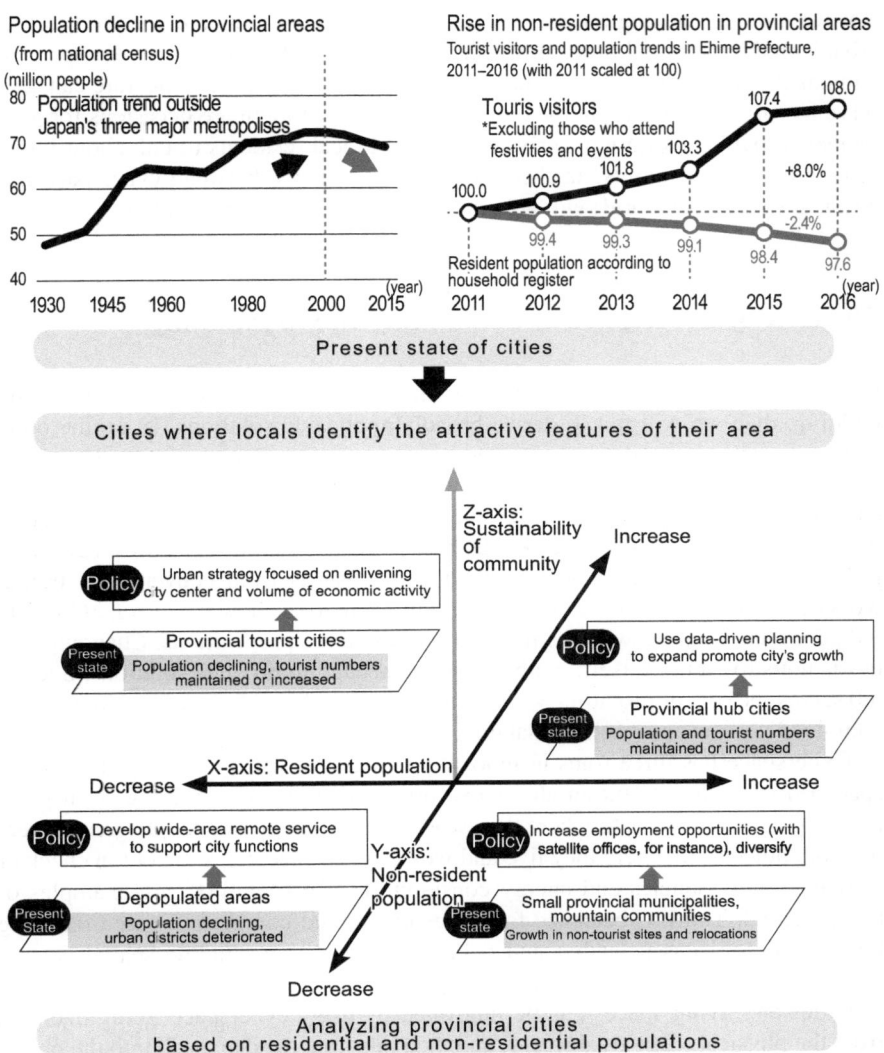

Fig. 5.3 Paradigm shift in sustainability of provincial cities

Hence, the fourth section of this chapter explores the necessity and potential of data-driven local planning and its associated tools.

Local governments must work out how they will gather local data on the physical space (real world), such as data on roads, buildings, people flows, and traffic, and how they will address the technical challenges in the building of a cyberspace infrastructure. No less important, however, is the issue of how the local community will use the cyberspace architecture. If provincial municipalities can establish a method for gathering Big Data on people flows, vehicular traffic, and the like, it would mean that we are approaching an age where data imagery systems are used to recreate urban conditions in real time. However, knowledge will go to waste if this process remains the sole preserve of a handful of experts. Cyberspace architecture can only serve its role of contributing to community building if there are occasions for governmental/public institutions, private firms, and local residents to talk about future visions of the community. In this way, the convergence between cyberspace and physical space (real world) will help create a truly citizen-led community.

The Role of Cyberspace in Community-Based Planning

If we think of the three propositions as logical extensions of traditional community building, then we will fail to derive broadly applicable solutions. If, on the other hand, we attempt to address these propositions using the Society 5.0 methodology, which emphasizes cyber-physical convergence, it becomes clear that we must change the traditional approach to community building. Once we replace values (the prevailing values that have held us back), create new values (draw on accumulated knowledge to create new values that release us from conventional frameworks), and revive abandoned values, we will begin to see how the vision of Society 5.0 goes hand in glove with the resolution of the three propositions. Changing values is not sufficient in itself; methodology is needed to produce solutions. Society 5.0 serves the role of providing broadly applicable tasks with a methodology that is combined with an approach to changing values.

Cyberspace has three roles in innovating cities and living spaces. First, cyberspace offers each resident an alternative office environment. In doing so, it gives people greater choice as to where they work. Second, in giving people greater freedom of choice as to where they live and work, cyberspace helps to hold in check or to reduce inefficiencies and energy consumption. BEMS/CEMS are examples of this potential. Third, cyberspace facilitates citizen-led community-based planning by allowing citizens to gather and collate Big Data (e.g., mobile spatial data or people-flow data) so as to share or evaluate future visions.

Cities and living spaces can be innovated by using cyberspace to digitize data from the physical space (real world) and then integrate the data back into the physical space. There is still some way to go until we see the full practical application of this process. In Sects. 5.2–5.4 below, we will explore the approaches and directions for technology development that facilitate innovating cities and living spaces in relation to each of the above three propositions.

5.2   Building a Habitat to Support the 100-Year Life

Society 5.0 and Habitat Design

The theme discussed in this section is how to design habitats to address the problem of the shrinking and aging population.

First, we should clarify the relationship between Society 5.0 and habitat design.

Human habitats are cultivated by two major forces: the market and government; however, these two forces alone can no longer cultivate the habitats adequately, so a third force is necessary. This is where Society 5.0 comes in: its role here will empower local communities/civil society to take responsibility for their habitats and thus ensure that habitats develop in accordance with the will of the people.

The development will be an inevitable outcome of three factors.

The first factor is that we are moving beyond the materially overabundant society.

Nowadays, the desire to acquire things as one's own and to experience things in one's own private space has been satiated. On the other hand, there is a mounting desire for (or a dissatisfaction with) experiences in public spaces, which one cannot acquire by one's own efforts. To give an example, many Japanese people lived in very cramped housing, often dubbed "rabbit hutches," in the 1980s, but living spaces have now become adequately spacious. Nowadays, the sources of people's frustration are things like the inadequacies of street environments, spaces for pedestrians, and public transport environment, and the lack (or poor quality) of places where people can gather and interact, such as shopping spaces, cafes, cultural facilities, and parks. However, when it comes to public products, such as public space designs and public policy drafting, the market principle does not apply. We therefore need a separate approach—namely citizen-led governance. Thus, Society 5.0 has the task of embedding into society a co-creation process, one that ensures that the will of the people is incorporated into the design and management of public products, including public space designs.

The second factor is that action on climate change can no longer be put off.

The ultimate public space is the planet Earth, and safeguarding our terrestrial home is a matter of paramount urgency. The governance of human activity and habitats (living spaces) can no longer be entrusted to markets and government alone; citizens must take the initiative and manage their habitats autonomously and systematically.

The third factor is that global digital networks and Big Data are enabling us to visually model complex living and activity spaces and to manage them instantaneously and organically. These technologies are also enabling the will of the people to be consolidated in real time.

Thanks to these three factors, human activities and living environments can now reflect the will of the people. In other words, Society 5.0, by deploying ICTs, AI, Big Data, and the like, opens up the possibility of managing the activities of civil society, private business, and government together with human habitats in an effective, autonomous manner that aligns with the will of the people.

The 100-Year Life: The Problem of the Shrinking and Aging Population

Japan has the highest rate of aging in the world. By 2050, almost 40% of the population will be aged 65 or older. People speak of the "ultra-aging society," a society where the elderly account for over a third of the population. As alarming as this may sound, insofar as we are heading toward an age of 100-year life spans, a graying population is only a natural outcome of longer life spans, and one that would happen even without a falling birthrate. Thus we really should not be so alarmed and dismayed at the prospect of this eventuality. After all, the prospect of living a fulfilling life until the age of 100 is something humanity has dreamed of since time immemorial. Thus, the high rate of aging should not, in and of itself, be a cause for chagrin—quite the opposite.

The real problem is how society can shoulder the growing number of people in need of care. As of 2010, there were 20 people of working age for every person with age-related care dependency. Assuming that the onset ratio of age-related care dependency remains at the 2010 level, there will be only ten working-age people for every care-dependent person in 2030, and this will decrease to five by 2060. This is indeed an alarming prospect. Long-term care insurance-based services have expanded, but family members care for dependents in two-thirds of cases, and few neighborhoods have access to round-the-clock in-home services. As family members become increasingly exhausted with their care burden, many elderly are forced to care for other elderly persons, many are forced to give up their careers, and some even start abusing their dependents.

Accordingly, the habitat design for the age of 100-year life spans must aim to achieve the three objectives shown in Fig. 5.4.

Supporting Autonomy in the Activities of Daily Living

What kinds of environments reduce the risk of becoming dependent on care? As of 2013, the following are the three biggest causes of care dependency:

1. The first is cardiovascular disease (stroke and heart disease), which explains 25% of cases.
2. The second biggest cause is motor impairment (fractures, joint disorder), which explains 21% of cases.
3. The third biggest cause is cognitive impairment, which explains 16% of cases.

It is now understood that cognitive impairment is related to cardiovascular disease.

Cardiovascular disease and cognitive impairment are lifestyle diseases, while fractures and joint disorders are the result of muscle degeneration, osteoporosis, and

(1) Maximize healthy and independent life spans: Minimize number of care-dependent persons and duration of care dependency	Living environments that encourage elderly residents to adopt healthy habits (healthy diet, mental/physical exercise, rehabilitation) and reduce risks of accident
	Local social environments that encourage elderly residents to engage socially in the local community, instead of confining themselves indoors.
(2) Local social environments that empower elderly residents to live independently in their homes even when they grow physically frail	Integrated local care systems and supportive environments that empower care-dependent elderly to live independently in their homes
	Public institutions, residents, and private firms collaborate in monitoring and assisting elderly residents (either self-help or public-help alone is no longer sufficient)
	Develop techniques that support independence (rehabilitation) and effectively prevent care dependency from worsening
(3) Increase working population	A society where older people can remain in employment: A society without a mandatory retirement age and one that accommodates a broad mix of workstyles and lifestyles
	A society of lifelong learning, where higher education or vocational training can be undertaken at any age and as many times as desired
	Universal design in living and work environments, whereby employees can contribute even if they experience disability
	Action against the falling birthrate
	Strategy to mitigate burdens of raising a family: As well as easing financial burdens (e.g., mortgage, schooling), aim to help parents balance career with raising a family (work-life balance) and ease the strains of child care.
	Strategy to accommodate immigrants: Multicultural society

Fig. 5.4 Three objectives for creating a vibrant society in a time of ultra-depopulation

deficiencies in the living environment. As such, the risk factors of these conditions can be minimized by better diet, exercise, rest, and living environments.

A well-designed living environment will reduce the risk of accidents occurring in the first place, and in cases where the person has suffered an accident-caused paralysis or other motor loss, such a living environment will support independent living through assistive technology. Notably, unforeseen accidents are the sixth biggest cause of death in Japan, and 75% of such deaths occur in the home.

Fewer than 4000 deaths in Japan are related to traffic accidents, while as many as 15,000 occur in the home—as a result of either falling or drowning in the bath. Even when domestic accidents are not fatal, the resulting fracture often means that the person becomes dependent on care (this is especially true among women). Strokes often occur while the person is in the toilet or bath, which is sometimes attributable to heat shock response. Accordingly, efforts to reduce the dependency risk factors in the living environment should focus on home installation and facilities that avert the risk of dizziness-triggered falls or heat shock responses. Another effective strategy is to introduce an AI-based system that can detect an emergency, such as a fall or cerebral infarction, and then summon an ambulance. This system would be particularly desirable in the case of cerebral infarction, as treating the condition within a few hours can significantly reduce sequela.

The Importance of Supportive Social Environments

Elderly people may recognize how important a healthy diet and exercise are in maintaining their activities of daily living. However, if they live alone, they may struggle to sustain healthy behaviors. Indeed, it is hard to sustain exercise regimens on your own and people tend to have poorer meals when they live alone. Rather than confining themselves to their homes, single-living elderly people should spend time outdoors socializing and dining with others, as such behavior is essential to their mental and physical health. This is starkly illustrated by a well-known statistic: compared to those who go out every day, people who go out only once a week are 4 times more likely to experience gait abnormality and 3.5 times more likely to sustain cognitive impairment.

Thus, when it comes to prolonging healthy life spans, in addition to (1) a healthy diet, exercise, and adequate rest, and (2) a safe living environment, a third component is necessary: (3) a local social environment that encourages elderly people to spend time outdoors engaging socially with the local community.

WHO's Healthy Aging Policy

The World Health Organization advocates a framework for preventing a decline in elderly people's capacities through local supportive environments. As Fig. 5.5 shows (World Health Organization 2015), the focus of public-health action changes depending on the phase of age-related decline, of which there are three. During the high and stable capacity phase, the focus is on promoting capacity-enhancing behaviors. During the declining capacity phase, the focus extends to reversing or slowing declines in capacity and supporting capacity-enhancing behaviors. Finally, during the significant loss of capacity phase, the focus extends to managing advanced chronic conditions and ensuring a dignified later life. Public-health action during this final phase also focuses upon removing barriers to participation. To this end, WHO advocates assistive devices such as wheelchairs, walking aids, and robotic assistance that provide frail individuals with as much independence as possible.

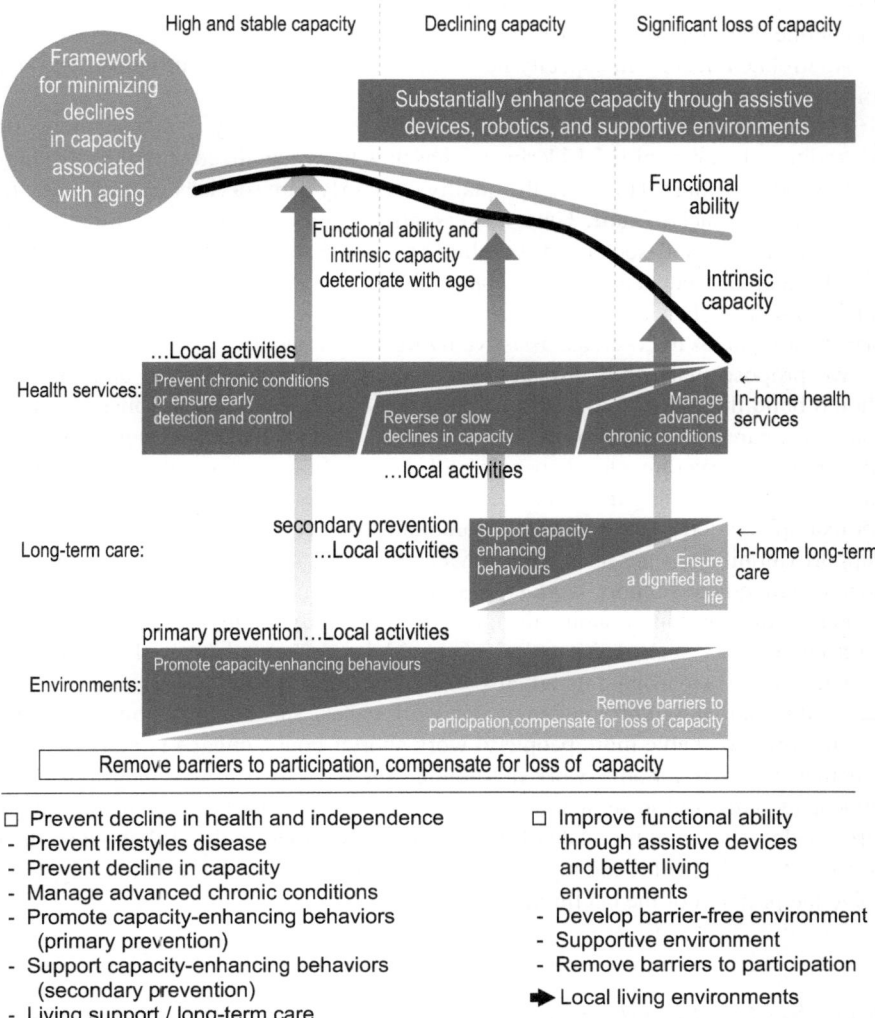

Fig. 5.5 WHO's healthy aging strategy. Source: WHO "Summary: World report on ageing and health" (2015)

Assisted Living Environments

The habitats that we should aim for are what WHO (in the above framework) refers to as "supportive environments"—environments that support elderly people's independent living and that encourage them to engage with their local communities. In this book, we describe this concept with the term "assistive living environments" and define these environments as follows.

Age-related decline in capacity necessitates care in six areas: mobility, bathing, toileting, cognition, sleep, and mealtimes.

Elderly people in Japan were once cared for exclusively by family members or by traditional communities. Elderly care has now been socialized with the introduction of social welfare. However, the quality of elderly care services is limited due to cost issues and lack of staff. Thus, it is essential to minimize the care burdens upon family members and society by building assistive living environments, in which elderly people's independence is supported by ICTs, AI, and robotic technology. Such ambient cyber infrastructure[2] will enable public institutions, residents, and private companies to co-create assistive living environments.

We propose that cyber infrastructure should be the foundations and that we should build three local infrastructural tiers upon these foundations: physical environments, care service environments, and social environments. Through such multilayered infrastructure, elderly people will access a full spectrum of care to support their independent living in the six areas mentioned above. The support will include, for example, elderly monitoring and communication services (cognition), mobility support (mobility), long-term care services (bathing, toileting, sleep, mealtimes), social engagement support, work style support, and childcare support.

Part of the mission of habitat design is to enable individuals to balance childcare with careers. As such, habitat design should be concerned not only with local (neighborhood) environments but also with social and spatial designs at a city and national level that will promote choice in work styles. Specifically, habitats must be reconfigured to enable more people to work shorter hours, engage in work-sharing schemes, work from home, work in satellite offices, or live close to their jobs. This radical reconfiguration in spatial and travel structure must be implemented on a large scale—that is, at the scale of the world's largest metropolis, Tokyo—and then rolled out to other cities and regions of Japan. Without doing so, we will struggle to solve the problem of the declining birthrate.

First, Set Out the Objective

In the previous section, we presented a rough framework. Of the support services in this framework, H-UTokyo Lab. has decided to focus first on developing elderly monitoring and communication services.

Specifically, we are developing ambient intelligence that analyzes imagery and audio data to make an informed decision on the user's status and provide cognitive and communication support/guidance infrastructure (from the user's perspective, the infrastructure resembles a robotic pet).

---

[2]Ambient cyber infrastructure is a network of sensors, robotics, and information devices that envelops the environments of human activity. Furnished with this infrastructure, a habitat can function as a robot monitoring and supporting human environments and human activity.

H-UTokyo Lab. (which commenced its work in 2016) decided that in its first 3 years of operation ending in December 2019, it will aim to develop a prototype of an in-home elderly monitoring system.

Objective

When elderly people suffer accidents in the home (such as falls or bathing accidents) or experience a stroke or cerebral infarction, the speed and effectiveness of the response have a critical impact on determining survival and prognosis. Moreover, a system that can effectively monitor elderly people who live alone (or who are alone during the day) is desperately needed to improve the QoL of caregiving family members and to prevent situations where family members give up their careers to care for the dependent. Thus, as a first step to supporting assistive habitats, where inhabitants can dwell in peace of mind, we will develop a prototype of an AI-based elderly monitoring system that can detect accidents or the onset of life-threatening conditions and then respond appropriately (such as by alerting emergency services). The system will be installed in the homes of elderly people who live alone (or who are alone during the day), in assisted living residences, and in care facilities.

Requirements

1. Must detect falls using imagery and audio data, and then alert emergency services.
2. Must use non-tactile sensing to detect the onset of cerebral or cardiac infarctions and then alert emergency services.
3. Must speak with the user so as to prevent false alarms.
4. During conversations with the user, the user's characteristics must be taken into account (this is particularly important given that the system is intended for elderly users). For example, the user's diction might be unclear if they remove false teeth at bedtime or bath time, and the user might also be hard of hearing. The system must also be able to distinguish the user's voice from other voices, such as those coming from the television or radio (perhaps the system can accomplish this by spatially mapping voices or by analyzing the characteristics of the voices).
5. Must detect accidents or onsets in noisy environments such as bathrooms and toilets (the user will be awake and active in these environments, so the system must function even with frequent calls).
6. The system will consist of an array of cameras, microphones, and speakers installed throughout the house, but the main interface, with which the user will converse, will be robotic pets (ideally, like cuddly toys) such as dogs, cats, and birds. For now, these can be immobile. In the bathroom, the interface can be a small LCD.
7. When sending an alert, the system must send imagery and audio data to the designated emergency contact (in general, this means connecting the home with the emergency contact by videophone).

8. Must send emergency alerts, manage the home environment, make calls, and make inquiries as instructed by the user (in general, this means having a smart speaker that can make videophone calls).

5.3   Carbon-Free Society: "Energy" × "Life" Management

A Masochistic and Non-masochistic Approach to Energy Saving

It has become customary to set room temperature for cooling to 28°C in Japan. In old days, the standard temperature setting was 26°C. The reason it increased by two points was because of Cool Biz, one of the campaigns in Japan to lower greenhouse gas emissions mandated under the Kyoto Protocol. In addition, although it was an emergency, the impact of energy saving from the Great East Japan Earthquake is also large. However, many people have now grown weary of such energy saving. The Cool Biz is not a problem. The 28°C is a problem because it is not suitable for office work. It is an uncomfortable temperature and it also results in lost productivity. There has never been any academic support for such a masochistic approach to energy saving. The recommended temperature is around 26°C in Japan: 28°C is the tolerance threshold.

So how can we save energy without having to suffer discomfort? One thing we can do is to cut wasteful use. We could, for example, turn off air conditioners in vacant rooms. This is an obvious action, and one that many people already practice. However, there is another way that might be less obvious: optimize operation of the building's air-conditioning system, lighting system, and so on. How to set room temperature is also one of optimization, but there is a plethora of other set points for control in the systems. With the right combination of set points, the same room temperature can be realized with less energy.

You may think, "Are these set points not optimized at the building design stage?" but they cannot be considered fully during the design process. Moreover, an air-conditioning system is built to order as well as a building. It is constructed by combining various devices, so there are generally some faults in the system. A troubleshooting process will always be necessary to iron out these faults. Therefore, optimizing of system operation should be conducted after building completion. However, that is not carried out in most buildings in practice. On the contrary, they do not even know what kind of operation of air-conditioning system they have currently. It might not be a completely fair comparison but, for example, in the automobile industry, there is a system to retain cars' performances continuously by periodic inspection and maintenance. However, there is no equivalent servicing for buildings once they have been completed and handed over to the owner (see Fig. 5.6). The potential energy savings from optimizing system operation are greater than we may imagine.

The "energy saving in operation" that engineers talk about refers to the depiction above, but when owners hear such a phrase, what occurs to them is a masochistic form of energy saving—i.e., enduring 28°C environments. Engineers try to plug this gap in understanding, but it is unreasonable as the concept is difficult to grasp without in-depth knowledge of how systems operate. This is not a recent issue but a big

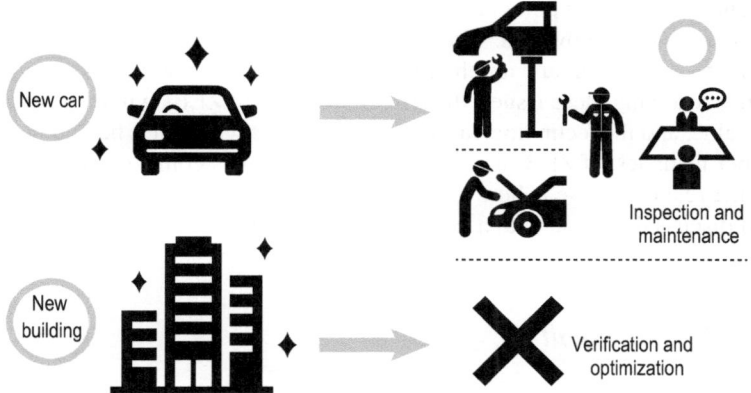

Fig. 5.6 There are scarcely any established business models for servicing completed buildings

barrier that has been around for a long time. Engineers probably lacked the motivation to break through this barrier because there were few business prospects. However, it is no longer acceptable to leave the barrier as it is.

Decarbonizing Existing Building Stock

Why can we no longer accept leaving the barrier as it is? The main reason is climate change. The Paris Protocol, effective since November 2016, indicated a change in mind-set: the signatories agreed to shift their sights from lower emissions to going carbon free, or "decarbonizing." Having signed the Paris Protocol, Japan is committed to a 26% reduction of greenhouse gas emissions compared to the 2013 level by 2030, and commercial and residential sectors are supposed to cut around 40%, respectively. This goal requires actions on the supply side, such as rolling out renewable energy, making thermal power stations more efficient, restarting nuclear power stations, and actions on the demand side, such as thorough energy saving in buildings. It is not only Japan that is focusing its efforts on commercial and residential sectors; this is the main focus among all developed countries.

According to building stock statistics and a building construction survey report of the Ministry of Land, Infrastructure, Transport and Tourism's Policy Bureau, as of 2015, there was a total of 1836 million square meters floor area of existing nonresidential buildings, and a total of 51 million square meters of floor area in new buildings. In the same year, there was a total of 5530 million square meters floor area of existing residential buildings, and a total of 79 million square meters floor area of new buildings. In other words, the floor areas of new buildings are only 2.8% of existing nonresidential buildings and 1.4% of existing residential buildings. There is little prospect that these percentages will rise in the future. These facts starkly illustrate the importance of decarbonizing existing building stock.

Of course, it is necessary to improve energy performance of new buildings. Japan is making advances in introducing zero-energy building (ZEB); these buildings are important from the viewpoint of not only decarbonization but also real estate value and technological innovation. However, the following issues remain: Are ZEBs being used as designed, and are their performances being maintained or improved? Without addressing these issues, the real estate value of ZEBs will overestimate the actual value, and the technological progress for ZEBs will be left behind because of no proper feedback. If ZEBs are also constructed, they become existing buildings.

So, the real challenge is how to decarbonize existing building stock. Hereunder, we will discuss this issue with a focus on management.

Energy Management

Figure 5.7 shows the 10-year trend in an office building's energy consumption from its completion (starting in 2004). A 40% cut in energy consumption was achieved over this 10-year period. Even if one discounts the decrease that was attributable to the Great East Japan Earthquake 2011, there was still a decrease of about 20%.

This was accomplished by improving energy performance through regularly monitoring the state of heat sources, air-conditioning, and lighting systems using hourly or minute-by-minute data, and identifying the optimal operations. Similar decreases in energy consumption have been achieved in buildings where there was such regular verification and optimization. This process is a part of "Commissioning (Cx)." Cx is a broad concept and is adapted through a building's life cycle. Cx is increasingly being applied to buildings, but it is still not common to do so in Japan. Here, we refer to the process mentioned above as Cx.

There are three major problems in the Cx. These problems represent three impediments to energy saving at the operation stage. The first problem is that human resources with specialized knowledge and technology to carry out the Cx are limited. This problem is related to building to order, and the complexity and the technical sophistication of the systems. There is little prospect of having a Cx professional for each building. Yet buildings have control/operation rooms, do they not? Yes, but most of these do nothing more than monitor the building systems. The second problem concerns the data processing. Cx deals with complex data sets with several thousand to several tens of thousands of data at increments of several minutes. Currently, these data are largely processed manually, meaning that the process is very inefficient and lacks a real-time element. As in the case of Fig. 5.7, a significant cut in energy use might be accomplished, but if it takes 10 years to do so, much energy and money during the period would be wasted. Air-conditioning systems operate differently depending on whether they are being used to heat or to cool, the way a building is used, as well as the meteorological conditions that change year by year to which the building is exposed. As such, it is probably to be expected that energy-saving efforts will take time. Still, 10 years is a long time to wait.

How can these two problems be solved? One solution is to have the limited number of Cx professionals conduct remote commissioning of multiple buildings. Another solution is to streamline and automate the Cx process using information technologies,

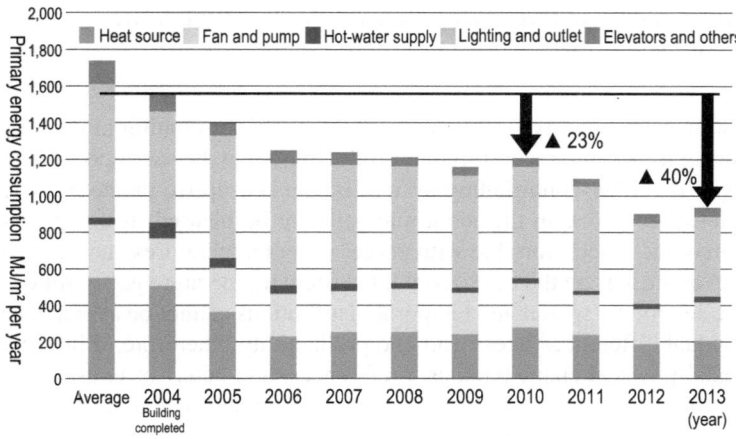

Fig. 5.7 Ten-year trend in an office building's energy consumption

such as AI, so as to accomplish energy savings within a short period after the building completion. As with our medical checks, continuous energy management is crucial for Cx where building system operations should be diagnosed and any problems should be rectified so as to restore the system to healthy operation. Thus, an energy management method for reducing system energy consumption with the Cx streamlined, automated, and conducted remotely is necessary. In 2016, Google reported that it had used DeepMind's AI system to cut energy consumption in its data centers by 40%. This is encouraging news, but coordinated and remote energy management is all the more challenging when it comes to large commercial buildings (especially multi-tenant buildings), which compared to data centers have many more variables to consider.

The third problem concerns something mentioned earlier: there is no established business model for managing the operation of building systems. This problem is related not just to technology but also to social institutions, business customs, and ethical concerns. In fact, building owners see little to be gained from energy-saving efforts in terms of cost-effectiveness. Although it does depend on the contract, owners of multi-tenant commercial buildings tend to be indifferent to energy efficiency, because the tenants typically pay lighting and heating costs at higher-than-usual energy unit prices, which include miscellaneous expenses. People used to speak of energy efficiency simply in terms of cost-effectiveness, saying that more efficient energy use will lead to lower heating and lighting expenses, but building owners do not seek to increase their profits through energy savings. To incentivize owners to take action in energy management, it is necessary to link energy management with other approaches. Accordingly, instead of evaluating energy-saving efforts based on short-term cost-effectiveness, we should use a longer term metric—namely, social return on investment (SROI), the social return in this case being the contribution to the prevention of global warming. SROI is closely related to the nonfinancial factors of environment, social, and governance (ESG), which have swiftly gained traction since the Paris Protocol. Thus, the future of energy management depends on whether Japan's building owners will identify the value in pursuing SROI.

Linking Energy Management with Life Management ("Energy" × "Life")

If your strategy for energy savings involves using the air-conditioning system sparingly during the summer, comfort and productivity may be sacrificed. On the other hand, even if the air-conditioning system is cranked up (to create a cooler room temperature), everyone in the room may still not be pleased. It is a problem that women may feel uncomfortable with cooler room temperatures and need to keep a blanket over the legs all the time. Even if the room temperature is controlled according to the set points, spatial and temporal distributions cannot be avoided, and there are individual differences in comfort and productivity. Therefore, if information on individuals' physical characteristics, preferences, and behavior is used to increase the adaptability of the environmental condition to the individual, these individual's quality of life (QoL) will improve. This is what we call life management.

This life management allows the comfort and productivity of individuals to be increased much more. This may have a positive effect on the real estate value for building owners. Greater productivity will lead to higher profit from limited work hours—something that will prove very attractive to tenants. An office's personnel expenses are said to be a hundred times its energy costs; comfortable and productive office environments will increase cost-effectiveness, which should be an attractive proposition to building owners.

Ostensibly, energy management and life management are independent of each other. However, insofar as life management is supposed to improve QoL, individuals' adaptability needs to be enhanced by unevenly distributing spatial and temporal environment conditions. It is necessary to measure QoL in a certain spatial location and time, and environmental control has to be changed from conventional common space to individuals to make the environmental condition uneven. Once the environmental condition has been distributed in this way, the environmental loads (in the case of cooling, the amount of heat removed from the room) will change accordingly, so energy savings may be achieved by cutting the loads. For both energy management and life management to function optimally, they must work in sync with each other.

Thus, the coupling of energy and life management work will produce a synergy that optimizes both energy efficiency and workers' QoL and, in turn, create economic benefits. The conventional (and poor) approaches to energy saving have undermined QoL and impeded economic activity. However, by pursuing energy management alongside life management, one can create an ongoing decarbonization cycle in the existing building stock. In other words, it is possible to innovate energy saving together with QoL (see Fig. 5.8).

To ensure that this vision of coupled management becomes a viable business model at the operational stage, it will be essential to make building owners realize that energy management will attract ESG investment and that life management will attract both ESG investment and tenants.

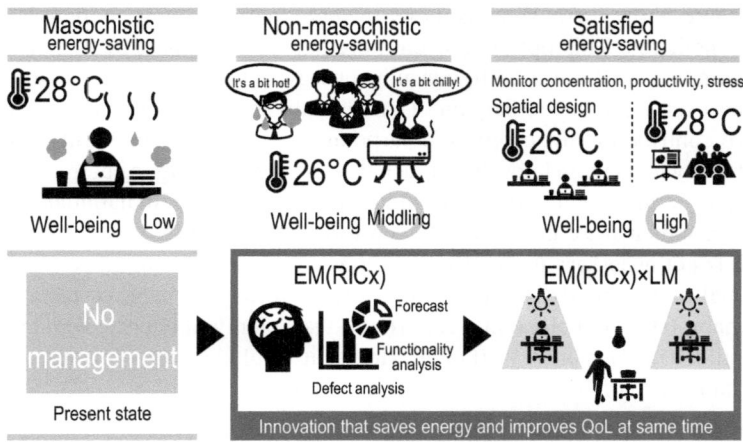

EM : Energy Management LM : Life Management RICx : Remote, IT-based commissioning (Cx)

Fig. 5.8 Masochistic energy saving, non-masochistic energy saving, and satisfied energy saving

The next section aims to give readers a better overall idea of "energy" × "life" management by outlining the concept through the lens of life management.

How Business Customs Affect Energy Use

If you look around the lecture rooms of a university campus, you will usually find a room with a single occupant who is busy studying, taking advantage of the room's bright lighting. The atmosphere of the room is conducive to concentrated study, so the student can get plenty of work done there. If it is the middle of summer or winter, the student will surely feel more comfortable if the room is air-conditioned. Regardless of how that student feels, it is obvious that air-conditioning this room, which is vacant but for that one student, is not an efficient way to use energy. However, the student is probably not concerned with this fact. Such lack of awareness is problematic. Is there no way that we could inform this student of how much energy his/her activity is consuming? Can we not convince him/her to continue his/her studies in a different location? As things stand, if a member of the teaching staff or the porter admonished the student, he/she would probably walk out in a huff.

Surely, many students have found that they get much more work done if they study in a café with a laptop, compared to when they study at home. Likewise, a growing number of white-collar workers are becoming "digital nomads" and carrying out their work in cafés. One must pay for the price of a coffee to "rent" a coffee space, but this poses no problem for the digital nomads. The return this investment promises—the comfortable space and productivity boost—far offsets the price of a coffee. It is also an energy-efficient way of working, because the air-conditioning is being shared with others. What is more, the café makes money, so there is an economic benefit too. This option is even more appealing if the café is a short walk from home; you will not need to consume energy in traveling to the café, and will get some exercise into the bargain. The students and digital nomads using these cafés are contributing to society to a surprising extent, albeit unwittingly. Can we not find a way to let them know how much they are contributing to energy efficiency?

Studying alone in a classroom and studying in a café are both habitual behaviors. Both environments are conducive to comfort and productivity. In these environments it is not necessary to make a conscious effort to study; in other words, there is minimal mental burden—one instinctively performs the habitual behavior. The former habit (studying alone in a classroom) uses plenty of energy, while the latter (studying in a café) is energy efficient. Hence, it would be preferable to replace the former habit with the latter.

Nudges

One often hears the term behavior change theory, which refers to theories and strategies concerning positive behaviors such as abstaining from smoking or alcohol or improving diet. The term became widely known, thanks to the application of behavioral therapy, a form of psychotherapy that sets out clinical interventions based on learning/behavior strategies. James Prochaska developed the transtheoretical model of behavior change (TTM). The TTM consists of five stages of behavior change and ten processes of change, of which five are empirical and five are behavioral. Prochaska argued that the key to successfully changing behavior is to strike the right balance between the positive and negative effects of the behavior change and to gain self-esteem. The physical and mental burdens associated with changing behavior must be offset. Stated differently, the change in behavior should lead to physical and emotional well-being and entail only minimal discomfort. The sense of satisfaction the person gains at changing their behavior motivates them to sustain the positive behavior.

A related term, one that often comes up in economics and marketing, is "nudges." A famous example of a nudge is the image of a housefly painted onto the urinals in men's public toilets. Notices imploring men not to make a mess during their micturition have met with only limited success. The image of a housefly, however, presents a "target" at which men will feel naturally inclined to aim. This ingenious leveraging of behavioral psychology has resulted in much cleaner toilets. Economist Richard Thaler made nudge theory widely known. Drawing on behavioral economics, Thaler outlined his ideas of nudges—ploys and strategies that "nudge" people into performing the desired behavior by their own volition rather than coercing them. People can be nudged by descriptive as well as injunctive norms. Descriptive norms concern perceptions of how people do in reality behave (whether rightly or wrongly). An example of a descriptive norm is the idea that if a crowd of people are crossing the street even though the light is red for them, you too can cross without fear. On the other hand, injunctive norms concern perceptions of how people should behave. You might be inclined to follow everyone else in crossing the street on a red, but you *should not* really do so. When it comes to energy saving, however, the reverse is true: if you notice that everyone else is saving energy, you might be inclined to follow suit, and indeed you *should* do so.

Life Management in Society 5.0

Society 5.0 remains somewhat ephemeral in terms of its key ideas, including the resolution of social issues with development, the supersmart society where all can live comfortable lives, and real-time exchanges between cyberspace and physical space (real world). However, efforts are being made to flesh out these ideas. Ever since the arrival of the Internet of Things, vast quantities of data from the physical space (real world) are sent to cyberspace, and from the data, new information is produced, which is then fed back instantaneously into the physical space (real world). Emotions such as discomfort and stress can be detected by sensors. Sensory perceptions and atmosphere can be extrapolated and communicated to others or relayed to remote locations. It will then be possible to forecast energy consumption and behavior. Society 5.0 will offer great value in terms of how information, sensory perceptions, and forecasts can be employed in real time. The ability to forecast and broadcast subjective human experience will help people adopt more pleasant behaviors; moreover, it will make it possible for the small choices that people make to generate sizable social value.

Let us consider free time as an example. Many workers in Japan skimp on break times to get more work done. This practice is problematic because it causes a buildup of fatigue. When you take breaks, your net working hours are less. Many Japanese workers avoid breaks for this reason—because they dread being thought of as lazy. Society 5.0 advocates well-timed, effective breaks. Fatigue and productivity can be continuously monitored. You could view real-time data that tells you how much employees could restore their productivity by taking a rest at a given time and how soon they might be able to complete their tasks. Stimuli, such as aromas, vibrations, sounds, or illumination, could be used to induce workers to take breaks. Then, objective data could be shown illustrating the uptake in performance following the rest compared to before. This data would help convince workers to rest—as they would see that taking a break is not synonymous with being lazy. More rest times will also help save energy because computers and lighting will be shut down during rest times. Workers will be able to see data informing them of how much they are contributing to energy efficiency.

That student in that classroom could avoid being scolded by doing his/her studies in a café instead.

5.4   Local Co-creation and Data-Driven Urban Planning

Why Data-Driven Urban Planning?

An important part of urban planning is transport. During the 1950s, Chicago became the first city in the world to introduce quantitative methods into its transport plan. As for Japan, the Hiroshima Major Metropolitan Area introduced the "person-trip" survey, leading to widespread use of the survey across the country in transport planning. Following Hiroshima's example, Japanese cities with populations of over 300,000 began to conduct the survey, sampling 3–5% of their populations. The survey data were used to guide and evaluate urban planning after the high economic growth period. The rise of this approach was underpinned by computer technology (see Fig. 5.9). Engineering workstations made it possible to run transport simulations that engineers could easily relate to. Economist Daniel McFadden used a behavior model to predict ridership demand on San Francisco's Bay Area Rapid Transit (BART) system. In 2000, McFadden won the Nobel Prize in Economics for his development of a model for analyzing discrete choice.

In actual urban traffic planning, you collect survey data on people's daily travel behavior, tally up volumes of the various travel paths, use the relevant statistical models to predict future travel behavior (such as generation/attraction, distribution, split, traffic assignment), and then formulate the transport plan accordingly. This process is based on mathematical models of transport that were developed 50 years ago. The data collection was paper based, but the data on people's daily transport behavior have grown increasingly sophisticated. Since the turn of the century,

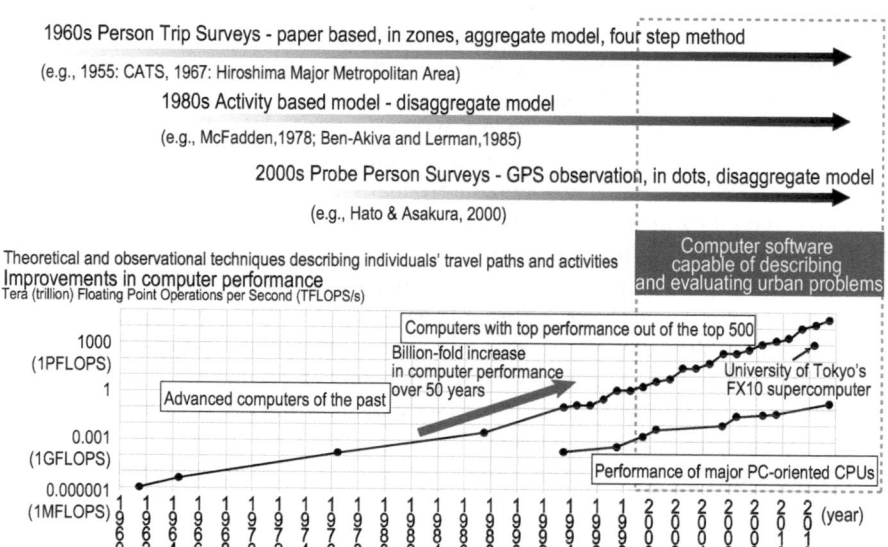

Fig. 5.9 Relationship between improvements in computer technology and urban planning

Japan has seen the emergence of the "probe person survey." Thanks to mobile communications, the probe person data are increasingly becoming directly available to urban planners. Indeed, we are close to the age of data-driven planning. Yet, so far these new methods have failed to gain traction in urban planning. Why?

Urban Planners Lose in a Lawsuit

In June 1989, an environmental organization called the Sierra Club Legal Defense Fund sued the Metropolitan Transportation Commission (MTC) of San Francisco. Why? The MTC had developed a model in 1977–1978, but it had failed to calculate the precise equilibrium values between the sub-models owing to the limitations of computer software at the time. The MTC argued that the construction of highways would ease congestion and improve the environment. This argument was premised upon the following causal chain: highway construction → highway capacity increases → highway speed improvements → lower emissions. The Sierra Club's retort was that the MTC's model failed to adequately account for the induced travel that resulted from congestion being eased.

What was the result of the trial? The MTC did not lose the case as such (the court accepted the plan to construct the highways), but the judge ordered the commission to make several revisions to the plan, effectively forcing it to develop new transport planning techniques. The defendant nominated individuals to serve as technical experts in the case. The plaintiff, being the world's largest environmental organization, did so too. The court appointed one of the nominees as a technical expert and, based on his insights, found that the MTC's transport-demand estimates were flawed.

Transport and urban planning often become subject to litigation. The reason is that urban plans restrict people's constitutionally enshrined land-use rights and affect their lives over a long period in the name of the public interest. It is now clear that public will not be convinced of the merits of public projects based on the actors who implement them or the procedures under which they are implemented. Accordingly, advocates of public projects must find ways to justify the projects on more rational grounds—i.e., they must proffer rational reasons justifying the public interest nature of the project. Accordingly, advocates must use data to adequately justify the project. Insofar as urban planning is a public act, the planners must be accountable to the public. Of course, in the interest of data privacy, the planners must not use personal data recklessly. However, when the public does not see the data, they may end up accepting a cobbled-together plan without that plan undergoing any improvements and, as a result, the public's freedoms may end up being significantly curtailed. Hence, urban planning must be data driven so that the public can understand which groups might be inconvenienced, when they may be inconvenienced, and how.

Examples of Data-Driven Planning

At present, it is not easy to pin down the meaning of data-driven planning. Arguably, conventional urban planning could be defined as data driven in that it used data (namely the "person-trip" survey data). The issue does not concern the technique then. Rather, it concerns whether the plans are scrutinized and debated with reference to quantitative data. If you live in a city with a population of over 300,000, you could check to see how many years it has been since the last "person-trip" survey was conducted. If the survey was conducted more than 15 years ago, then you should see this as a red flag. It indicates that the city should revamp its traffic system urgently, reconfiguring the travel paths in districts with aging populations and redistributing road routes to foster communities within walking distance. Because there is now the demand that new initiatives accord with the circumstances of local communities, we should be wary of plans or policies that are not shaped by comprehensive and quantitative survey data.

Figure 5.10 shows an example of smart planning in Kobe's Koikawa-suji street. When Kobe City revamped a district in the city center, it took a data-driven approach: it used Wi-Fi data and data from a "probe person survey," in which subjects carry GPS mobile phones and make online travel diaries so that their travel paths are traceable online. As the figure shows, Kobe City presented simulated outcomes of its project to expand sidewalks and create pedestrian-only environments, including number of visitors, city center stay time, city center walking distances, distances from entrances, and stay time in given locations. The data illustrated that creating pedestrian-only environments would be very impactful and enliven the area. The stakeholders (e.g., a local retailer association, Kobe City, transport companies, the police) discussed the plan with reference to the numerical data.

	Present state produced	Sidewalk extension	Enhancement/ pedestrian-only environments
Number of visitors	2.21	2.10 (-5.0%)	2.11 (-4.5%)
Minutes spent in city center	267.8	279.4 (+4.3%)	271.6 (+1.4%)
Meters walked in city center	432.6	441.7 (+2.1%)	540.5 (+24.9%)
Maximum number of meters walked from entrance point	200.5	189.5 (-5.5%)	225.0 (+12.2%)
Minutes spent in each area	124.2	130.4 (+5.0%)	125.5 (+1.0%)

Fig. 5.10 KPIs for project to redistribute pedestrian space in the Motomachi area

The Future of Cities

The city is humankind's greatest invention. Cities house enormous economic potential as illustrated by the fact that many companies claim that business-to-business innovation is only possible in Tokyo. On the other hand, in *Urbanism as a Way of Life* (Wirth 1938), Louis Wirth defined cities as places that require urban planning to control the size, density, and heterogeneity of the population aggregate. These urban issues continue to pose challenges to urban planners in the twenty-first century. There is little sign of progress in solving issues such as energy, migration, and congestion; on the contrary, these issues have grown in severity. Many creative millennials will be living the 100-year life, but these individuals tend to opt for more fluid lifestyles. Increasingly, people select the city rather than the city selecting them. Cities are great for doing things on a large scale and doing things efficiently, but urban environments often lack a human touch—that sense of inspiration and sensuality. Humans are social animals, and we seek environments that enliven us sensually and provide emotionally stimulating interpersonal encounters. What then is the ideal urban environment to aim for?

Figure 5.11 shows a photo of an urban design school that forms part of the Urban Design Center Matsuyama (UDCM). It also shows a green space called Minna no Hiroba; both are undertakings by Matsuyama City, Ehime Prefecture. Matsuyama City was full of car parks, so we decided to construct a green space, Minna no Hiroba, in a car park a little away from the commercial district in the city center. We ripped up the asphalt and found an old well. After running tests on the water, we formed a knoll with a fountain in the center for the public to enjoy. The fountain is now a spot where children gather. We also set up the abovementioned design school in a boarded-up shop across from Minna no Hiroba. The design school serves as a forum for discussing community-building projects that use Matsuyama's resources, including camellias and "Iyo-Kasuri" fabrics.

Fig. 5.11 Minna no Hiroba (Matsuyama, left) high line (New York, right)

There is increasing interest in the benefits of creating open spaces in urban environments that would otherwise be saturated with drab, utilitarian features. There is also interest in how higher quality spatial designs can generate novel ideas regarding urban stock such as car parks, roads, and elevated walkways.

An example of the latter is New York's High Line (the photo on the right in Fig. 5.11). The High Line is a linear green space running along what was once an elevated railway line. The structure was about to be destroyed, but Joshua David and Robert Hammond led a campaign to save the High Line and renovate it as a relaxed parkland. Their organization keeps the site clean and cultivates gardens with many different kinds of vegetation. How can we shape our urban environments? As democracy withers, cities around the world need more than ever to have people-centric urban planning that is grounded in data and behavior.

Transcending City Boundaries

The use of data does not guarantee that the city will improve. Urban planners need to have data on their side. Cities face a mountain of problems. A major task in the twentieth century was how to introduce automobiles into cities. In the twenty-first century, urban planners must work to find ways to introduce automated driving into cities, which requires hitherto unseen urban designs. When it comes to Society 5.0, we must work out what such a society should look like. Local communities, public

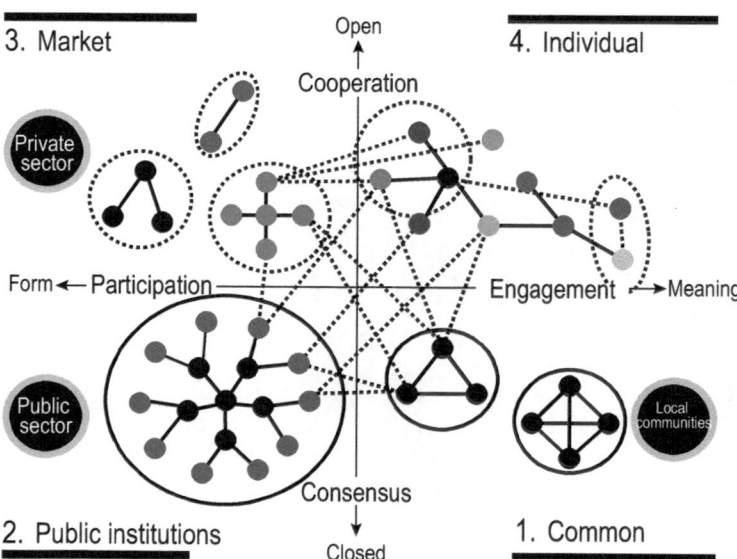

Fig. 5.12 Data-driven planning

institutions, markets, and individuals can each play a leading role in mapping out this society. However, each society has weaknesses. We must support and augment each society by engaging, cooperating, and sharing ideas (see Fig. 5.12). Cities will lose their spark if they are too homogenous. Likewise they will fail if they are too disorganized and incoherent. The boundaries of cities should be flexibly defined so as to ensure human and urban creativity as well as stability and security. Data-driven planning is nothing more than a means to prompt dialog on urban boundaries. Without this dialog, urban boundaries will become rigid. This is something that we must avoid.

References

Aguilera A, Wenglenski S, Proulhac L (2009) Employment suburbanisation, reverse commuting and travel behaviour by residents of the central city in the Paris metropolitan area. Transp Res A Policy Pract 43(7):685–691

New York Times (2008) According to a New York Times article dated 24 Feb 2008, the reverse commuting rate rose by 12% from 2000 to 2005, https://www.nytimes.com/2008/02/24/nyre-gion/nyregionspecial2/24Rreverse.html. Accessed 5 June 2019

Wirth L (1938) Urbanism as a way of life. Am J Sociol 44(1):1–24

World Health Organization (2015) Summary: world report on ageing and health. http://apps.who.int/iris/bitstream/handle/10665/186468/WHO_FWC_ALC_15.01_eng.pdf. Accessed 5 June 2019

Chapter 6
From Monetary to Nonmonetary Society

Atsushi Deguchi, Shinji Kajitani, Takahiro Nakajima, Hiroshi Ohashi, and Tsutomu Watanabe

Abstract As a consequence of the digital revolution, we predict the dynamic change of our daily lives and consuming activities, and moreover we have foresight on the possible impact to our economic systems and human relations.

Section 6.2 discusses the impact of unbundled innovation to the economy and the factors underpinning the unbundled economic activities, and approaches the advantages and issues of the digital platform to be installed in the economic system of a data-driven society. Section 6.3 approaches the issues of the cashless society from the economic aspect of a data-driven society. It points out two types of possible issues: pricing the priceless information and managing personal data without anonymity in the cashless society, which the digital currency enables to realize.

Sections 6.4 and 6.5 are the philosophical approaches to the humanity and human wealth to be aimed as the goals of Society 5.0. Section 6.4 suggests the development of current sharing economy method and the economic paradigm shifts: from the conventional economy based on private ownership to the new economy based on

The original version of this chapter was revised: This book was inadvertently published with the incorrect license type CC BY 4.0 and the Open Access License has been amended throughout the book to the correct license type CC-BY-NC-ND. The correction to this chapter is available at https://doi.org/10.1007/978-981-15-2989-4_9

A. Deguchi (✉)
Department of Socio-Cultural Environmental Studies, Graduate School of Frontier Sciences, The University of Tokyo, Tokyo, Japan
e-mail: deguchi@edu.k.u-tokyo.ac.jp

S. Kajitani
Department of Interdisciplinary Cultural Studies, Graduate School of Arts and Sciences, The University of Tokyo, Tokyo, Japan
e-mail: kajitani@fusehime.c.u-tokyo.ac.jp

T. Nakajima
Institute for Advanced Studies on Asia, The University of Tokyo, Tokyo, Japan
e-mail: nakajima@ioc.u-tokyo.ac.jp

H. Ohashi
Graduate School of Public Policy, The University of Tokyo, Tokyo, Japan

Graduate School of Economics, The University of Tokyo, Tokyo, Japan
e-mail: ohashi@e.u-tokyo.ac.jp

T. Watanabe
Graduate School of Economics, The University of Tokyo, Tokyo, Japan

collaborative commons, and from the society with the conventional value for owner-ship to the society with the new value for usage. Section 6.5 approaches the image of future society to be aimed by Society 5.0 from the view of humanity and philoso-phy. It suggests that Society 5.0 should innovate the capitalism for transforming from material based to human based together with the growth of human capability, which can be called as a society for "human co-becoming."

Keywords Cashless society · Digital economy platform · Nonmonetary society · Post-capitalism society · Sharing economy

6.1 Data-Driven and Nonmonetary Society

So far, this book has outlined the concept and nomenclature of Society 5.0 and dis-cussed the approaches to and future directions of technological development based on such a concept. So far, Society 5.0 has been discussed chiefly from an engineer-ing perspective. This chapter takes its perspectives from economics and humanities. From these perspectives, it discusses the future of the data-driven society—the kind of society that Society 5.0 espouses—how we can grasp/embrace such a society, how viable it is, and what issues will emerge.

We have already explained the background and the purports for why Society 5.0 was outlined in the government's 2016 fifth Science and Technology Basic Plan. To reiterate, this was related to the fact that Japan and other developed nations have reached a major turning point technologically and socioeconomically. This situation is all the more obvious now compared to 10 years ago. Ten years ago, smartphones had not caught on, nor had things like car sharing and blockchain. Over the past 10 years, Apple, Amazon, Google, and Facebook have achieved a meteoric rise to world dominance.

Meanwhile, China has made startling progress in going cashless, and companies like Alibaba and Baidu have become major players. In large Chinese cities today, you can buy a drink from a vending machine without cash, and neither do you need cash to take a taxi or purchase something from a stall. Many Chinese people have not used any cash for over a year. In a cashless society, goods are paid for in cyber-space, and everyone's purchase history is stored as Big Data.

Once you buy a book on Amazon, your inbox will receive a succession of rec-ommendations (stating "you may also like the following items") on further items based on what you purchased or viewed. These recommendations are an example of what cyberspace systems accomplished with the use of AI to analyze Big Data (customer purchase history) and then actively prompt the customer to make further purchases. In a data-driven society, the results of AI analysis are deployed in psy-chological ploys to induce particular types of human behavior. The businesses are concerned with behavioral economics. Thus, it is not surprising if someone buys some tens of books in the space of a few months that they would otherwise not have bought. A concern of the data-driven society is that large businesses will make customer purchase histories snowball into a vast data reserve, and then monopolize all the massive profits that this data can yield. Such an eventuality is likely in China, with its market of a billion people. The winners in such a data-driven society will be the companies who gather Big Data.

How then can companies in smaller countries like Germany and Japan ride with or resist this tide? Given that apps and smartphones play a leading role in the data-driven society, perhaps the future of these companies lies in developing constituent technologies for these things. The rapid proliferation of data services means that it is harder to succeed in business solely on manufacturing prowess. The world is placing more value on data and less on manufacturing technologies. It was against this backdrop that Society 5.0 was proposed as a vision of future scientific and techno-logical progress as well as a vision of a future society. In this respect, Society 5.0 is not unconcerned with the global proliferation of the data-driven society.

With the global spread of capitalism exacerbating economic and regional inequali-ties, many fear that the data-driven society will lead to further social divisions and disempowerment. To address these concerns, the Comprehensive Strategy on Science, Technology, and Innovation for 2017 states: "Society 5.0, the vision of future society tow [sic] which the Fifth Basic Plan proposes that we should aspire, will be a human-centered society that, through the high degree of merging between cyberspace and physical space, will be able to balance economic advancement with the resolution of social problems by providing goods and services that granularly address manifold latent needs regardless of locale, age, sex, or language to ensure that all citizens can lead high-quality lives full of comfort and vitality." Whereas the present thrust of capi-talism is one that exacerbates division, Society 5.0 offers an alternative form of capital-ism, one in which scientific and technological progress transforms regional disparities into opportunities for each local region to promote its unique qualities and transforms diverse preferences and lifestyles into an inclusive, accommodating society.

The paradigm shift we aim for in Society 5.0 is one of values; we seek a shift to a people-centric society, one that is inclusive of different communities and individu-als and is not overly focused on economics. Economics measures things by mone-tary value, but people-centric values—in particular, QoL—cannot always be measured in monetary terms. The role of nonmonetary values is therefore a central concern in the discourse on Society 5.0.

There is an alternative view. A web browser is a key example of a nonmonetary service. We search the Web every day to find information that we personally value, and yet we do not pay service fees directly for these searches. Users can come up with business ideas for monetizing the information they gather from these non-payable online services.

Section 6.2 of this chapter discusses the potential for developing digital plat-forms in Society 5.0. Section 6.3 discusses the role of cash in the data-driven soci-ety, where individuals' purchase histories are archived. Section 6.4 discusses the meaning of wealth in the sharing economy. Finally, Section 6.5 outlines "human co-becoming," a concept of human independence in a data-driven society.

Discourse on the future of the data-driven society will increasingly concern the question of how monetary and nonmonetary economies will conflict or coexist. I hope that this chapter will prompt readers to consider this issue.

6.2 Digital Platforms in Society 5.0

Society 5.0 represents the next step in our socioeconomic evolution, the previous steps being hunter-gatherer (Society 0.1), agrarian (Society 0.2), industrial (Society 0.3), and information (Society 0.4). Each of these steps forward was the result of what Bresnahan and Trajtenberg (1995) called "general-purpose technologies," which provide an engine of growth that transforms existing social structures. Each time an old system was replaced with a new one, our life and work styles were transformed accordingly, as were our values and ways of thinking.

In the agrarian era, farming was a general-purpose technology. Hunter-gatherer communities became sedentary and started rearing livestock and producing crops. Village communities began to emerge as the basic social unit, giving rise to a land-based economy. Meanwhile, societies became stratified into rulers and ruled. Steam power began to develop in the early seventeenth century and it eventually became a new general-purpose technology, which enabled dramatic increases in productivity and thus sparked the shift from agrarian to industrial society. In that industrial era, populations gradually shifted away from rural communities and into urban districts, resulting in a large-scale clustering of labor into cities. Around this time, Japan started shifting away from the traditional social stratification known as the "four categories of the people" (gentry scholars, peasant farmers, artisans, and merchants).

Our generation has lived in the information society. One of the general-purpose technologies in this era is IT, including computer technology and satellites. Television, newspapers, and other mass media have narrowed the information gaps between different regions, and there are now much greater flows of people, goods, and money. However, there are also stark regional disparities; many local communities are disappearing, while in the cities, people are much more likely to interact with strangers in their workplaces and living spaces. Since the 1990s, Japan has been on a privatization path amid the tide of structural reforms and regulatory easing, and this has raised the question of how to maintain nonprofitable public services.

In Society 5.0, the general-purpose technologies will be ones that monitor and analyze in real time and optimally manage society as a whole, in other words, technologies that manage human behavior as well as energy and transport infrastructure. Society 5.0 will have cyber-physical systems, thanks to the ability to use advanced AI systems to analyze unstructured Big Data gathered by the Internet, sensors, and digital technology. This section explores this coming era from an economics perspective with respect to Society 5.0.

Unbundled Innovation

As the wry adage goes, "it's difficult to make predictions, especially about the future." The more cautious economists are, the less inclined they are to forecast the future. British economist John Maynard Keynes must have been very bold therefore when he penned the 1930 article "Economic Possibilities for our Grandchildren," in which he forecast how the economy would look in 2030 (Keynes 2010). Keynes predicted that "the standard of life in progressive countries one hundred years hence will be between four and eight times as high as it is" and that there would be a "15-hour work week." He also predicted that his generation's grandchildren would see an end to the economic problems that have bedeviled humankind since time immemorial, causing us to fight over basic resources. According to Keynes, "there will be great changes in the code of morals" and "the love of money as a possession … will be recognized for what it is, a somewhat disgusting morbidity."

Almost 90 years have passed since Keynes made his predictions. Our standard of life, as measured by GDP per capita, is ten times higher than it was in 1930, exceeding Keynes' prediction. Keynes would, however, have been disappointed in other respects: we have made little progress in labor distribution, while job insecurity, economic inequality, and poverty have grown even worse. We may not have reached what Keynes called "our destination of economic bliss," but in the case of Japan at least, what we desire today is markedly different from what our forebears desired in the 1950s, when the must-have items were the "Holy Trinity" of the black-and-white television, the washing machine, and the refrigerator. Today, consumers have shifted their interest from tangible goods to intangible services, and their desires are to experience something rather than to own something. This would explain why we are seeing increasing demand for peer-to-peer services (shared economy) and virtual/augmented reality when it comes to cars and accommodation.

Innovation concerns technology, but it also leads to changes in people's behavior. Many past examples of innovation led to "unbundling." The rise of the sharing economy, for instance, has decoupled use from ownership. Likewise, mobile phones have unbundled communication from fixed locations (landlines). Similarly, recordable TVs have unbundled the experience of viewing a TV program from the timeslot in which the program was broadcast, and massive open online courses (MOOCs) have unbundled education from the classroom.

Such unbundling not only affects the demand side but also shapes supply. Whereas suppliers have outsourced manufacturing overseas to minimize costs, the rise of 3D printing and other forms of advanced manufacturing are creating new possibilities for factories and R&D sites to optimize their operations without needing to worry all the time about production costs. Unbundling is also changing the way we work. It has created new forms of employment, leading to a broader notion of work (for example, telecommuting is now seen as an acceptable way of working) and opening up possibilities for freelance work, something that was not part of the conventional notion of work. Labor services are nowadays provided in an environment where work times and work locations no longer necessarily overlap, which makes it necessary to develop institutions that allow for more organizational flexibility.

The Economic Factors Underpinning Unbundling

What are the social and economic effects of unbundling? Perhaps a useful way of approaching this question is to consider the economic nature of digital services. There are three aspects to consider. The first concerns cost structure. Building a digital service platform entails hefty fixed costs for things like setting up a user interface. On the other hand, the marginal cost of reproducing services is negligible.

The second aspect is industrial structure. Traditionally, it is the service provider who bears the fixed cost, so the service provider must have sufficient financial clout. However, the rise of digital platforms has changed the situation. These platforms match supply with demand in real time, enabling services that were traditionally bundled in terms of time, space, and organization to be delivered unbundled.[1] In other words, the platform provider is unbundled from the service provider. Since the service provider bears no fixed cost and only minimal marginal cost, mass customization is possible. The platform provider on the other hand, in hosting the unbundled array of services on its platform, must exercise financial clout and work hard to recover the fixed cost.

The third aspect is demand structure. If many users flock to a platform, the platform will also attract a large number of service providers along with their various services. In a competitive market, this network effect (when the economic value of something increases in proportion to the demand for it) will lead to the more popular and successful service providers dominating platforms. Once monopolized by a service provider, a platform will serve as the service provider's business base, creating an economic ecosystem.

Open Community Platforms

The shift from the industrial to the information society was accompanied by an increase in people flows. In cities especially, much of the social and economic interactions are between strangers. By contrast, traditional communities would have long-standing neighborly networks based on which the community members would barter with each other and owe each other favors.[2] However, as it became increasingly common for transactions to be between strangers in communities with no hierarchical power relations, it became difficult to form long-standing trust relationships. Accordingly, money became a much more convenient means to pay for things. The majority of transactions then started being conducted in a market space, where people were free to enter and exit as they pleased, as opposed to within insular communities. Under these circumstances, it made sense for money to circulate widely.

---

[1] In the field of social infrastructure, this concept corresponds to publicly built but privately operated facilities as well as to the separation of infrastructure from operation.

[2] Even today, vast quantities of rice in Japan are given free to intellectuals and relatives.

Price is a critical piece of information that needs to be communicated to buyers. Price, in a matching process, is determined by the market mechanism, or what Adam Smith called the "God's invisible hand." However, just because things are priced does not necessarily mean that they will be traded efficiently. If buyers cannot easily observe the quality, then according to Gresham's law, which states that "bad drives out good," low-quality goods will drive out high-quality ones (Akerlof 1978; Ohashi 2017). Oftentimes, it is necessary to create an alternative mechanism to communicate the value of quality. An example is a certification system, in which a designated organization certifies a product or service, assuring buyers of its quality. Such a mechanism is essentially an attempt to recreate the kind of trust-based transactions within traditional communities. In a traditional community, sellers have an incentive to maintain quality because if they sell poor quality, they are penalized in some way.

In Society 5.0, the market mechanism should be more sophisticated and able to correct faults in the market. Big Data gathered by the Internet, sensors, and digital technology will be subject to sophisticated AI-based analysis, enabling economic transactions to be conducted across digital platforms that communicate various information, not only price. Some elements of this system are already here. Uber, for instance, provides both driver and rider information and lets riders rate their drivers. In Society 5.0, these platforms will allow the best of both worlds—a borderless market in which one can enter and exit as one pleases and, at the same time, a community-based market that gives buyers a range of information other than just price. The idea of an open community might once have seemed like an oxymoron, but digital platforms, in matching supply with demand, do indeed combine openness and community.

The Advantages and Problems of Digital Platforms

The open communities that digital platforms will serve an indispensable market function in Society 5.0's trading. These platforms facilitate trade by indicating nonmonetary information as well as monetary price. This information empowers buyers to make informed choices about what to purchase, and the culmination of these consumer choices will encourage businesses to develop more creative products and services to compete.

Markets should be fair, but they should be so a priori (at the outset) as opposed to a posteriori (in outcome). Some businesses will succumb to competition and be forced out of the market. One occasionally hears the argument that markets should be a level playing field a posteriori, but we must remember that if we let every competitor be a winner, there will be no incentive to enhance quality or efficiency, and so buyers will lose out. So although we cannot make digital platforms fair a priori and a posteriori, we must also bear in mind two competition-related issues (Ohashi 2018).

First, when it comes to public services that are essential in our lives, such as infrastructure, we must reproduce the system of mutual supplementation that existed in communities. Take, for example, the privatization of infrastructure. The government is pursuing a plan to entrust the management of infrastructure such as waterworks and roads to private operators as part of a structural reform and regulatory easing project intended to encourage creative innovation in the private sector. Traditionally, the whole infrastructure was maintained through cross-subsidization; profitable infrastructure propped up unprofitable infrastructure. However, if profitable infrastructure is in private hands, the survival of unprofitable infrastructure becomes doubtful. As public services become increasingly marketized, services with doubtful profitability may be shed. We need a system that distinguishes between those services that should emphasize profit and those that should prioritize the public good over profit.

The second point is that we must address the information asymmetry in digital platforms. Austrian-born economist Friedrich Hayek saw markets as places for communicating information. Through market-determined pricing, participants' private information is shared on the market as public information, which allows the market to play a public role—that of balancing demand with supply. In so doing, the market stores public knowledge and becomes democratized.

However, digital platforms differ from Hayek's conception of the market in that the platform operator profits. There is considerable information asymmetry between the platform operator and the platform participants; the latter share their knowledge with the platform operator but not with each other. If the knowledge becomes a tool of the platform operator, then this nullifies the advantage of the participants possessing knowledge; consequently, the participants' services become commodified. This situation creates a profitability gap; platform operators achieve sustained profitability by gathering the knowledge and using it to make their operations more efficient, while the platform participants struggle to maintain profitability because their services are commodified. We are already witnessing these gaps growing at an alarming rate in digital platforms.

If the platform participants have the option of switching to an alternative platform operator, they may find a way to avert the commodification of their services. However, if there is a strong network effect, this will create the winner-takes-all situation described earlier, eliminating all but a few platform operators. This bottleneck will deprive the participants of choice.

The Consumers' Society 5.0

In *Future Shock*, futurists Alvin Toffler argued that economists are "conditioned to think in straight lines" and thus tend to see the future as a "straight-line projection of present trends" with no break from the past (Toffler 1984). This tendency has become all the stronger in today's society, which calls for evidence-based decisions and evaluations.[3]

---

[3] In Japan, for example, there is now evidence-based policy making right across government.

This section might not have added significantly more to Tofflers' critique, but it has discussed how digital platforms offer an advantage (in that they create open communities) and disadvantages (the bottleneck and information asymmetry) in the context of Society 5.0, a society that seeks to further promote human liberty.

To minimize the problems of digital platforms, we must find ways to restrict excessive cutthroat competition, and this can be achieved through the general-purpose technologies of Society 5.0 and the science underlying them. The general-purpose technologies are vulnerable to monopolization, so we will need social institutions that can prevent this risk. The EU's 2018 guidelines offer some suggestions to this end, in particular, the regulation on promoting fairness and transparency for business users of online intermediation services (The regulation on promoting fairness and transparency 2019). This regulation enshrines the principle of fairness in transactions between platform operators and related businesses. The fact that bottlenecks can so easily occur in the platforms makes it all the more necessary to ensure transparency and impartiality, the requisites for fairness. Only once this fairness is assured will platforms function properly as highly advanced markets, allowing buyers to thrive as "opportunity-creating" (Masuda 1989) entrepreneurs and setting the stage for Society 5.0.

6.3   Role of Cash in a Data-Driven Society

Two Ways of Going Cashless

Cash is the most essential infrastructure for underpinning people's economic activities. IT and the IoT transform cash in two main ways.

First, they make cash digital, where once it was physical. The expression "going cashless" usually refers to promoting monetary transactions through credit or debit cards or by other alternatives to handing over hard cash. As used here however, "going cashless" refers to the use of digital currency as an alternative to hard cash.

The Bank of Japan (BOJ), which is responsible for issuing the nation's banknotes, can track the circulation of each 10,000 yen banknote based on its serial numbers. The BOJ cannot, however, tell who currently holds the banknote or what it has been exchanged for and where. In this respect, hard cash has a very anonymous element. This anonymity is one of the defining features of hard cash, but it also represents a technical limitation. With digital currency, on the other hand, you can, at least in principle, trace who has the money and where it is being used.

In a data-driven society, the more data there are the better (as these data are the fuel that "drives" the society). Yet the cost to anonymity cannot be ignored. The key to making digital currency a success then is to address people's fears about losing their anonymity. This personal data issue is the most important issue to address when designing the data-driven society. As we see with the recent EU discourse on data portability, the debate over personal data boils down to the issue of who has the right of ownership over data such as one's purchase history. The anonymity of digital currency is an emblematic example of this issue.

The second kind of cash transformation concerns the proliferation of moneyless transactions. In other words, people are buying things without hard cash or electronic money. When we buy something, we usually pay for it in money. This payment provides a source of revenue to the seller. In this way, money becomes the economic lifeblood.

That does not mean, however, that money mediates all transactions. When parents prepare meals for their families, we do not expect the family members to pay money for the service. Moneyless transactions also prevailed in rural communities until fairly recently: farmers would distribute surplus crops to their neighbors, and neighbors would lend a hand with the farm work pro bono. There have been communities larger than families in which money did not mediate the members' relationships.

If paying for things with money is "monetary economics," then paying for things without money is "nonmonetary economics." Historically, nonmonetary economics prevailed, but monetary economics rapidly proliferated after the Industrial Revolution. Nowadays, we usually measure a country's economic well-being by the scale of its monetary economy, and disregard its nonmonetary economy. Hence, the nonmonetary economy is considered only minimally when calculating gross domestic product (GDP). The reason is that there is a tacit understanding that nonmonetary economy tends to be smaller relative to the monetary economy.

However, this situation has recently started to change. Technological innovation is driving the proliferation of nonmonetary economies. The world is increasingly going cashless. "Going cashless" might not be an ideal term, but it does usually refer to the proliferation of nonmonetary economies. Wikipedia is an example of this trend. It was not so long ago that each family kept large encyclopedias, such as the Encyclopaedia Britannica, on their bookshelves. These encyclopedias of course had to be paid for, and they were by no means cheap. Adults and children alike would look up facts in these encyclopedias. Nowadays, we use Wikipedia instead. Wikipedia is very convenient; one can look up something easily and the articles are updated frequently. Moreover, it is free to use. Fewer people use traditional encyclopedias, and unsurprisingly, Encyclopaedia Britannica's sales are flagging.

To pay for an encyclopedia with cash is an example of monetary economics. To look up something on Wikipedia for free represents nonmonetary economics. Thus, an economic activity that was once monetarized has become non-monetarized.

Consider another example. Figure 6.1 shows the rate of increase in the number of photos taken throughout the world. The rate begins to rise gently in the latter half of the twentieth century, after which it skyrockets. This development illustrates a change in the economic significance of photography. In the past, pictures were captured on film and then developed and printed. The process was accompanied by payable services and products provided by the manufacturers of cameras and film, as well as the shops that developed and printed the images. Nowadays, people take snaps on their smartphones and upload the images onto social media; they do not require the photos to be developed or printed. Camera manufacturers have no input in the activity. Hence, companies such as Kodak are feeling the pinch. As this example illustrates, we can see that monetary economics is the preserve of traditional companies that fail to ride the wave of technological innovation, while nonmonetary

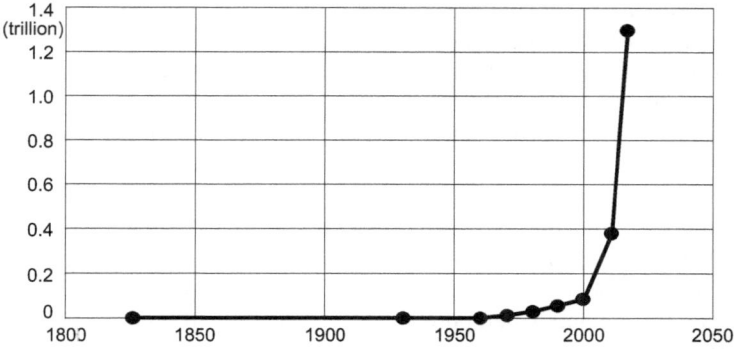

Fig. 6.1 Number of photographs taken in a year. Source: https://digital-photography-school.com/history-photography/

economics is the preserve of companies that achieve success underpinned by technological innovation.

What are the implications of the rise of nonmonetary economies? In monetary economies, the pricing of goods and services acts as a signal that contributes to a positive loop, in which the more people want the goods or services the more they are produced. This is called the pricing mechanism. Products that are more popular (desired by more people) will fetch higher prices. The producers of such high-price products are then motivated to increase their supply, as doing so will earn them profits. The increased production will give more consumers a chance to buy the product and thus spur more consumption.

In nonmonetary economies, the reverse is true. Because there is no pricing, the producers are unsure at what volume to produce the products. Consequently, the supply can be low even when the demand is high. Another difference is that although monetary economies can be measured using economic metrics such as GDP, these metrics have little use in nonmonetary economies because the activities therein are not calculated in monetary terms. Thus, a new kind of system is necessary to measure the vibrancy of nonmonetary economies.

In the following section, we consider the backgrounds to these two forms of cashlessness and their significance.

The Society That Digital Currency Enables

Consider first the benefits of converting hard cash into digital currency. The first benefit is that the costs associated with cash transactions are cut and the Japanese economy is made more competitive. Cash is primarily used in banking and circulation. According to Mizuho Bank's estimates, banks typically spend around 2 trillion yen a year on managing accounts and maintaining ATMs. Mizuho also estimates that retailers and restaurants spend around 6 trillion yen a year processing cash transactions, for a total of 8 trillion yen per year. When consumption tax is taken into account, as much as 4% is spent every year on cash processing. One can see how this 4% would eat into the profits of banks and retailers; it should also be obvious that using a more efficient financial resource would boost Japan's economic competitiveness.

The second benefit is that digital currency will enhance Japan's security. With cashless transactions, national borders become irrelevant. Many retailers in Japan have introduced cashless payment services, such as Alipay, in an effort to attract Chinese tourists into stores. Given the sheer popularity of Alipay in China, the service might one day catch on among the Japanese too. If it does, then it would mean that payments in Japan will be processed by a Chinese company, and the payment history data (where and which purchases were made) will go to China. This situation would threaten our economy, not to mention our national security. To offset such a risk, we should take the initiative in making our own cashless system.

Anonymity and Personal Data Management

What schemes are needed to digitize money? Two broad kinds of digital currency are required. The first is privately issued decentralized digital currency. The second is centralized digital currency issued by public institutions such as a central bank. An example of a decentralized digital currency is Bitcoin. Many other decentralized currencies exist, underscoring their considerable market potential. However, Bitcoin and its equivalents have so far been used primarily for investment; they are not widely used for payments. It is hard to imagine that these decentralized currencies will ever replace cash. We advocate the other kind of digital currency. Specifically, we believe in a centralized digital currency that people can trust because it is backed up by a public institution, such as the BOJ. We also believe that this currency should be stably tied to the yen at a one-to-one exchange rate. The confidence this currency commands will make it less costly than its decentralized counterparts.

The most prominent example of a digital currency managed by a central bank is e-krona, which is issued by Riksbank, the central bank of Sweden. This centralized digital currency would allow account holders to transfer their funds to each other digitally. People could use e-krona to pay for things in stores, send funds to each other, and split a restaurant bill. Both Krona banknotes and e-krona represent a claim on the central bank, so they share the same simple structure: payments are made by transferring them.

However, there are three problems with people holding accounts in a central bank. The first concerns anonymity. In the case of banknotes, the central bank cannot tell who is using them and where. However, when account holders use their accounts to transfer funds, the use of the money is plainly visible to the central bank. Some worry that central banks could maliciously exploit this information. Whether or not their fears are justified, most account holders would at least accept that the details of their transaction cannot be completely confidential. Until anonymity is ensured, digital currency will fail to gain traction.

The second problem is that if people can directly hold accounts in a central bank, this would put the central bank into competition with the private banks and their settlement accounts. Currently, there is a reasonable balance between the use of settlement accounts and the use of banknote cash, but if members of the public hold central bank accounts, this balance would be undermined.

The third problem is that broader access to central bank accounts may disincentivize innovation in the private sector. The only technologies that will see practical application will be those that align with the central bank's agenda. If private firms and banks come up with innovative ideas, they might end up being used by the central bank. This situation would inhibit innovation in digital currency. Already, the Japanese Government and the BOJ have a stranglehold over the issuance and the circulation of banknotes, leaving precious little room for private innovation. Digital currency should provide room for technological development; it should not reproduce this status quo.

As an alternative to holding accounts in the central bank, digital currency transfers could be between private bank accounts. This form of digital currency would not threaten private banks' settlement account businesses. It would also address the anonymity issue to some extent, as the central bank will not see the transfer details. These details would, however, be seen by the relevant private banks, so anonymity would be no stronger than it is in the present banking culture. This alternative is similar in many respects to Mizuho's J-Coin Pay and MUFG Bank's MUFG Coin. Both services are pegged to the yen at a one-to-one exchange rate.

To ensure that people trust digital currency as much as they trust their banknotes, the private banks must stipulate clear principles on how they will manage the funds in digital accounts. The most easy-to-understand example is full-reserve banking, the principle that banks should keep the amount of each depositor's funds in the central bank. Many other principles could be used besides this, but the point is to ensure confidence in the banking system that supplies the digital currency, and the central bank along with government (financial regulatory authorities) should play a role in establishing these principles. In other words, digital currency should be designed and provided through public–private partnerships (involving the government, the central bank, and private banks), and indeed such partnership is necessary.

Pricing the Priceless

Now let us consider the other kind of cashlessness—namely the increasing use of moneyless transactions (transactions that involve neither hard cash nor digital currency). Many e-services such as social media platforms and search engines have something important in common: they use nonmonetary pricing models. Twitter, Facebook, Wikipedia, and Google are all free to use. This is a boon to users but a bit of a headache for economists.

GDP is the key metric for measuring economy activity, but it only applies to transactions of payable goods and services. The metric cannot account for nonmonetary transactions. The popularity of Wikipedia cuts into encyclopedia sales, decreasing Encyclopaedia Britannica's contribution to GDP. Wikipedia itself contributes nothing to GDP because it is free to use. Consequently, GDP declines. The sluggish GDP in recent times reflects this phenomenon to some extent.

Some may wonder how Google and other e-commerce companies can earn so much and whether economists may have made a mistake somewhere. Google's turnover is indeed enormous and much of it comes from advertising. Google also earns profits from marketing users' data, such as their search histories. Although users pay no money for using Google, they do pay in other ways, such as putting up with advertisements. Users essentially barter for the service by offering to put up with the ads, and so no money changes hands.

Bartering means exchanging one thing for another of the same value without using money. A barter, though nonmonetary, could therefore be measured in monetary terms. Some have attempted to indicate the monetary value of e-services so as to quantify their economic value in terms of GDP. The estimates will naturally have their fair share of errors, but judging from the sets of estimates we have seen so far, e-services make only negligible contributions to GDP.

Why should this be the case? Perhaps there is no real bartering going on after all. Users for their part do indeed face the cost of putting up with ads, but how does this cost stack up against the economic value of Google's services? Google's chief economist Hal Varian estimated that Google has an economic impact of 150 billion dollars, significantly outstripping its 36 billion dollars in ad revenue. This estimate seems to suggest that Google is selling itself short; could it not be making a lot more money? It is doubtful that Google would willingly sell its services for a song, so somewhere along the line it must have failed to price its service at a level that reflects the extent to which users appreciate it.

A New Problem with Pricing the Priceless

We have just discussed the question of how to create an alternative pricing mechanism, but there are a host of other issues related to nonmonetary pricing models. One such issue is the divergence between production (e.g., GDP) and economic impact (user's satisfaction). In monetary economies, impact is generally tied to production, so it suffices to check the GDP. However, technological innovation has undermined this linkage, such that GDP can no longer be used as an indirect/alternative indicator of impact. Thus, we urgently require direct measures of economic impact.

Some have suggested using willingness to pay (WTP) or willingness to accept (WTA) as direct measures of economic impact. WTP describes the maximum amount of money that consumers would be willing to pay for a free product. WTA describes the minimum amount of money that consumers would be willing to accept to abandon a free product. Both WTP and WTA are gaged through consumer surveys.

To give an example of these consumer surveys, the team at the University of Tokyo's Watanabe Lab surveyed users of Line, a freeware messaging app. The average WTA (the minimum amount respondents would accept for abandoning Line) was, on the condition that the respondents' Line contacts continued to use the service, 4,070,000 yen per year. This finding suggests that a typical Line user values the service at 4 million yen. There is sizable interindividual variation, but the level of WTP and WTA remains fairly constant even when one discounts the larger responses. When this 4 million yen figure is multiplied by Line's extensive user base, said to be 70 million strong, it amounts to a massive sum indeed. If Line's economic impact is as massive as this, then its actual revenue is tiny by comparison.

Although WTP and WTA are effective measures of economic impact in theory, it may not be practically feasible to conduct the surveys on a scale sufficiently large enough to gage the overall state of a nonmonetary economy. There instead needs to be a technology that can measure nonmonetary activity granularly and frequently. Hitachi developed a system that uses sensors to measure happiness; something similar to this is needed for nonmonetary economies.

6.4 Private Ownership to Collaborative Commons: Wealth in a Postcapitalist Society

Envisaging a Future Society

What do you value in your life? What kind of life do you want? Each person has his/her own answer. Happiness and wealth are ultimately defined by the individual. At the same time, we all live amid the social circumstances of the day. Happiness and wealth are defined in the context of these circumstances.

How a person lives depends on how they interact with the society. Whether you go with the flow or swim against the tide, your life is a refraction of the social conditions of the time.

Society 5.0 is supposed to be different from the societies to date, but exactly what sort of society should it be? In Society 5.0, what will humans value, and what kind of happiness and wealth will they seek?

Society 5.0 is the vision of the future society outlined in the 2016 5th Science and Technology Basic Plan, which states, "(Society 5.0 is) so called to indicate the new society created by transformations led by scientific and technological innovation, after hunter-gatherer society, agricultural society, industrial society, and the information society" (Cabinet Office 2016a).

Society 5.0 remains a catchphrase with little in the way of concrete details. This is well illustrated by the fact that the term "Society 5.0" is not accompanied by a descriptor (such as *hunter-gatherer*, *agricultural*, *industrial*, and *information*). The Basic Plan itself concerns science and technology, and as such, it highlights ways society can use AI, IoT, nanotechnology, Big Data, and similar innovations.

Thus, the kind of society the Plan advocates is one in which production and sales are thoroughly streamlined through high-tech developments such as a "high degree of merging between cyberspace and physical space" and a "supersmart society" (Cabinet Office 2016b).

The Plan also defines the supersmart society as "a society that is capable of providing the necessary goods and services to those who need them at the required time and in just the right amount; a society that is able to respond precisely to a wide variety of social needs; a society in which all kinds of people can readily obtain high quality services, overcome differences of age, gender, region and language, and live vigorous and comfortable lives." However, there is precious little detail on how technological progress will usher in such an egalitarian society. Will technological progress naturally produce such a society by itself?

History is indeed replete with cases where new technology led to a new society. The invention of the printing press led to the proliferation of knowledge and had a critical impact on education. The proliferation of home appliances socially empowered many people, particularly women. Yet we should also remember that the social consequences of technological progress depend on how the technologies are used. Television and newspapers are used as channels of free expression in democratic societies, but as tools for propaganda and control in those that are totalitarian.

Insofar as Society 5.0 is a vision of a new society, its advocates must think about the shape of this future society. We must also understand how values may change; otherwise society might head down the wrong path, leading to chaos and suffering among people who struggle to adapt to changing times.

We must consider this issue in relation to capitalism, or to put it another way, in relation to monetary value. After all, capitalism is today a critical factor that shapes society most deeply and broadly. Should Society 5.0 be a logical extension of capitalism, or should it be a break from capitalism? We can consider this question by focusing on value and wealth.

What Is Wealth?

The most basic kind of value in a capitalist system is monetary value. In this respect, capitalism has made many societies rich.

From the time of Japan's high economic growth period until the 1980s, incomes rose and socioeconomic inequalities narrowed. This was a time when most Japanese people identified as middle class, as expressed in the slogan *ichiokusōchūryū* ("100 million middle class"). From the 1980s onward, capitalism widened socioeconomic inequalities both nationally and internationally, creating widespread poverty. Undeniably though, wealth has been maintained at a society-wide level. Even more importantly, there is a common society-wide understanding of the *meaning* of wealth, and society and individuals alike emphasize the importance of acquiring wealth according to this meaning.

So what is wealth? In the capitalist sense, wealth at a national level is expressed in GDP—the market (added) value of all goods and services. When a nation's GDP rises, it indicates economic growth and greater wealth. At an individual level, wealth increases when the person's wage increases. Both types of wealth are monetary. To obtain wealth, nations seek to increase their GDP, and individuals seek to increase their wages. In this way, capitalism relentlessly drives the pursuit of growth.

But as Tomas Sedlacek asked in *Economics of Good and Evil*, can we have capitalism without growth, and can we find a way to wealth without economic growth or higher wages (Sedlacek 2013)?

In the minds of some readers, these questions may have evoked the idea of abstaining from pleasures and leading a frugal existence of scrimping and saving. However, even today there is no scarcity of examples of wealth that cannot be measured by GDP or wage levels.

Each prefecture of Japan has monetary measures of wealth such as GDP and wage levels, which indicate how rich or poor that prefecture is. We tend to define regions as poor when they have low GDP and wage levels.

However, this does not necessarily mean that the inhabitants of these regions are poor; just as incomes in these regions may be lower than those in urban areas on an average, so too are the living costs (goods and housing). Some people would find life out in the sticks dull due to a lack of cultural and leisure activities (hence, there is an outflow of young people), but the countryside is not without its own kind of wealth: the pace of life is more relaxed, there is plenty of delicious and inexpensive products, and one can lead a healthier lifestyle. When it comes to education, rural areas face a disadvantage in that there are relatively few cram schools and activity clubs, but urban environments have high schooling costs, which can squeeze family budgets. Arguably, a price cannot be placed on raising a family amid the abundance of nature.

If you live in a rural community, you might have to lend a hand in community undertakings such as clearing land and festivities. Townies might regard these voluntary activities (or rather, duties) as burdensome obligations. These obligations do indeed put many people off from moving from the city to the country. Then again, in return for fulfilling these obligations, one can partake of mutual community assistance in its various forms.

I am not trying to say that life in the country is a rich life and that city living is a poorer life, but that monetary measures like GDP and incomes cannot simply be used to determine where life is rich or what kinds of lives are rich.

What then is nonmonetary wealth?

Monetary and Nonmonetary Wealth

In a capitalist system, wealth is market value, that is, exchange value in monetary terms. What is exchanged for money depends on what is traded on the market. Capitalism assumes that economies grow when there is a continuous increase in the range and scale of market trading. All goods and services traded on the market are monetarily valued, and the higher the value is, the larger the GDP will be.

The goods and services traded on the market are all desired by at least someone for some purpose, but not everything on the market is truly desirable or would contribute toward a richer life. Medicines represent an example of goods that do not make people richer. If many people have a certain disease, drugs that can prevent or treat this disease will sell in high volume, along with related products, and GDP will rise as a result. Another example is disposable products, which people use and then discard without a second thought. These products contribute to GDP in that they can be continually produced and consumed. They further contribute to GDP in that they lead to services related to the reuse or recycling of the disposed products.

Although these things contribute to GDP and incomes, they do not necessarily make people's lives richer. In some cases, they may even decrease QoL.

When we consider it, it becomes obvious that monetary value is disconnected from the richness of our lives, even if it partially overlaps. Indeed, much of our wealth cannot be measured in monetary terms.

The richness of our lives is a product of psycho-spiritual qualities such as camaraderie, affection, goodwill, sincerity, trust, serenity, and self-confidence. These things exist outside the market and they are unexchangeable. They have no monetary value.

That is not to say that they have nothing to do with money. Some argue that you cannot be happy without money and that money can buy love, and they are not completely wrong. If clothing and food are ample, then people understand ritual and moderation. When we lack the material necessities, we experience inner turmoil too.

The reason poverty is associated with a lack of monetary/exchange value is that when one cannot afford things monetarily, one often cannot acquire nonmonetary things either. Camaraderie and love, for instance, are not measured in monetary terms per se, but they may require the acquisition of things that are monetary.

So people who renounce all but the most basic material necessities in pursuit of an esthetic poverty and simplicity will not live a rich life, unless, that is, they really are able to live with only the bare necessities. Inasmuch as the nonmonetary psycho-spiritual qualities are not constituent parts of capitalist society, society will be less likely to define these things as valuable, even if these things do contribute to individuals' well-being.

"Use Value" Without "Exchange Value"

So psycho-spiritual qualities such as happiness, love, and trust have no intrinsic monetary exchange value in that they cannot themselves be exchanged on the market. There are also examples of things that once had monetary value and were once traded on the market (even today, they continue to be traded in part), and yet have all but lost their monetary value.

It is not that no one needs these things or that there is no need to use them. On the contrary, they are exceedingly valuable and they are in use. In other words, they have zero exchange value but paradoxically have a high use value. They are things that many of us use free of charge. Today, the world is awash with these things, and they are increasing in number.

What I am talking about are e-services, which use ICTs, and the Internet (which itself is free). Examples include freeware, email, message boards, Skype, Line, and Facebook. These e-services have become akin to social infrastructure: they are so valuable that we could scarcely live without them. Under market principles, these e-services are not exchangeable in and of themselves, but they underpin the very process of exchange.

The fact that e-services have use value without exchange value does not mean that they represent a rare exception or a fluke. According to Jeremy Rifkin, these e-services reflect an economic shift from capitalist markets to a Collaborative Commons (Rifkin 2014).

Rifkin argues that capitalism will, by historical necessity, lead to its own demise, giving way to a Collaborative Commons. In a capitalist system, Rifkin asserts, companies seek to increase their profits, and they do so through technical innovation and cost-cutting measures, which are designed to improve productivity and minimize marginal costs (production costs per unit). Those companies that accomplish this task effectively will gain the upper hand in a price war, allowing them to corner the market and nudge out their competitors. This process will create continued competition in price and quality (provided that the market is not monopolized by a single company or by a cartel). Sooner or later, the marginal costs will approach zero. Eventually, the products and services will become tantamount to free, and profits will also be erased.

According to Rifkin, this outcome is the final destination of free-market capitalism. Rifkin cites publishing as an example. Although the process will not occur for each and every book, and neither will the process occur at the same speed in each case, the digital publication of e-books will remove the costs of publishing itself while also making the content readable for free. Similarly, Skype allows free video calls. Education is another example; MOOCs and other kinds of online courses allow people anywhere in the world to access education services for free or at minimal cost. Likewise, many software programs can now be downloaded as freeware, whereas they once had a hefty price tag.

In energy too, the proliferation of small-scale renewables will lead to zero marginal costs. In addition, the rise of 3D printing and the arrival of free design software enable the creation of all manner of products in private homes or small production sites, as well as larger sites such as factories. Taken to the extreme, an individual might even be able to directly manufacture the products needed.

IoT—the online connectivity of tangible things (such as buildings, vehicles, home appliances, and manufactured goods)—allows us to understand where goods are in short supply and where they are in surplus, allowing us to efficiently fill in shortages. For example, Airbnb lets users exchange information on vacant rooms and to offer these rooms as lodgings. Uber facilitates car sharing in a similar manner. This peer-to-peer sharing extends to clothes and other daily necessities.

Some of these services are payable. It is not always clear when payable services will become free, but as Rifkin argues, the Collaborative Commons will only grow ever larger in the future. This trend is not necessarily at odds with capitalism. The Collaborative Commons is, in fact, supported by capitalism and it develops in tandem with it. In the course of this process, fewer goods and services will be exchanged, leading to a smaller GDP, but that does not mean that wealth declines. The question then is what does wealth mean in the context of this process and what changes will occur in the values underpinning such wealth?

Sharing as a New Value

Economies generally distinguish between exchange and use value, but in the Collaborative Commons, the value of products and services might not purely be their use value. When the marginal costs of a product become zero and the product becomes free to use, it will cease to be exchanged. That is not to say, however, that everyone will cease using or desiring the product. It is important to note that in these circumstances the product's use value will increase, not decrease. The reason why the product is not being exchanged is not because it lacks value; it is simply because the product is being shared.

In such circumstances, there is value that should be shared, and value that is generated from sharing. Conversely, market capitalism is premised on (private) ownership, and exchanges occur when there is a transfer in ownership rights. With sharing however, there is no such exchange. If everyone uses a product, it means that the product is shared. With such sharing, exchanged products and services will enter the market and gain monetary value. In this way, common value is a requisite to monetary value.

The Collaborative Commons will expand the bounds of common value, such that many products will be commonly accessed without anyone privately owning them. This situation will decrease market transactions, GDP, and incomes, but wealth will remain high.

The present capitalist society defines wealth as how much an individual privately owns. In the Collaborative Commons, wealth is measured by how much is shared. One can be rich without owning lots of things and without earning the money necessary to own lots of things—it is not necessary to be Mr. Moneybags to be rich.

It might be possible to quantify this new sense of wealth using ICTs and the IoT. Many of the technical innovations underpinning Society 5.0 are closely related to common value and the Collaborative Commons.

However, we cannot be so certain that sharing is correlated with happiness. Perhaps, the more one shares, the more stressed one becomes, as there will be more things to worry about. Higher amounts of sharing can also entail a greater amount of management, which could easily be used to justify surveillance and control by the powers that be. As we work to make Society 5.0 a reality, we must also address the question of how technology can overcome these dangers.

6.5 Society 5.0 and "Human Co-becoming"

What Is Society 5.0?

What kind of society does Society 5.0 aim to realize? Some would say that it is a society underpinned by technologies such as the IoT, Big Data, and AI that overwhelmingly exceed human abilities. Such a society might be utopian, but it could potentially be dystopian too. We can see Society 5.0 as a future utopia, in which we live comfortable and convenient lives, largely emancipated to a great extent from the need to work, while we can see this as dystopia—a society where humans are in fact controlled by technology, such that they have nothing meaningful to do but languish every day in utter boredom.

Whether utopia or dystopia, the dream (or nightmare) itself of a technologically advanced future society is not particularly new. Throughout the twentieth century we have attempted time and again to envisage such a futuristic society. If there is a new opportunity in the idea of Society 5.0, it would be relevant to rethink the way of living of humanity in a world where we are blessed (or controlled) by advanced technology.

The Modern Humanity and Capitalism Based upon Things

If we are to rethink what humanity is today, we have to interrogate the relationship between humanity and capitalism, the principle that has significantly regulated the contemporary world. Michel Foucault asked this question around half a century ago in 1966:

> As the archaeology of our thought easily shows, man is an invention of recent date. And one perhaps nearing its end.
>
> If those arrangements were to disappear as they appeared, if some event of which we can at the moment do no more than sense the possibility—without knowing either what its form will be or what it promises—were to cause them to crumble, as the ground of Classical thought did, at the end of the eighteenth century, then one can certainly wager that man would be erased, like a face drawn in sand at the edge of the sea. (Foucault, Michel, *The Order of Things: An Archaeology of Human Sciences*, London: Routledge, 2002, pp. 422)

The era of classical thought, which existed through the seventeenth and eighteenth centuries, gave way to the era of modernity. The era of modernity that existed in the nineteenth century was underpinned by the concept of "homme" that is "man" or "humanity." According to Foucault, this concept would come to an end in the twentieth century.

The development of capitalism is keenly connected to this shift of the eras and the concept of humanity. Adam Smith published *The Wealth of Nations* in 1776, heralding the arrival of modern capitalism and the modern concept of humanity. His major idea is as follows: in the era of classical thought, wealth was based on exchange of goods, while in the era of modernity, it was based on production of things by human labor. We could characterize this modern way of production as capitalism based upon things. Human labor produces things, from which wealth is derived.

The Consumption of Differences and the Rise of Capitalism Based upon Events

Foucault thought that this modern paradigm began to shift in the twentieth century. What happened in the twentieth century, particularly in the latter half? Capitalism shifted its focus from things to events. In other words, capitalism based upon events emerged. Capitalism went to handle information and happenings as events for its investment. Amid an ocean of information stirring up our desires and prepackaged happenings for our experience, we started consuming differences and were ourselves reduced into consumable differences. Modern "subjectivity"—which was never realized in its full meaning—was dissolved into pieces. Instead, humanity as difference or relationality appeared.

However, what we have to ask now is the question of what actually defines us as humanity. Humanity is something singular, which is irreducible either to consumption or to the order of the difference. We are urged to interrogate what humanity is after Foucault's criticism. At the same time, we must think how we can imagine the forthcoming capitalism after capitalism based upon things and events. Thus we can start sketching out the future society which the idea of Society 5.0 tries to elaborate.

"Human Capitalism" and "Human Becoming*"*

I would propose, as a hypothetical concept, an idea of "human capitalism." By using this concept, I am figuring out the humanity neither as laborer, nor as consumer, nor as humans as nodes of difference, but as *value*. Once advanced technology emancipates or deprives us from labor and consumption, what aspect of humanity will become the focal point of capitalism? I think that we need to reformulate capitalism so that it helps us create human value, rather than depriving us of it. To this purpose, it is inevitable to think what the ultimate value is for the humanity.

To put it bluntly, the value for humanity is the transformation of humans themselves. Influenced by modern economic discourses, we often think that value is something that we own as property. It does not work well, because it is just a conversion of the value of the commodities that we produce and consume into human value. We have to separate human value from property-based imagination.

Let us imagine once again what a future society empowered by advanced technology might look like. In this Society 5.0, what would we possess as values? Automated vehicles? Smart AI systems to provide the optimum solutions by analyzing Big Data? Or creativity of arts irreducible to the advanced technology? How do you think of it? It seems that these ideas of property-based values are too clichéd.

The twentieth-century imagination of the future society lacked a possibility that humans would fundamentally be transformed. Philosophically speaking, the idea that humans will be transformed equates to the idea of the human as *becoming* something human, as opposed to the Western traditional idea of the human as *being* or *having*. I propose to think of human becoming instead of human being by referring to Roger T. Ames (Ames 2010; Rosement Jr and Ames 2016). The word "capitalism" derives from the Latin *capitalis*, meaning "head," and a person's head is a matter of life and death. The future of capitalism will certainly be a matter of critical importance, determining the fate of human life and death.

Capability and Social Mobility

What will this critical matter be for us? The answer is simple: becoming *human*. We cannot become human by ourselves. It is only when others come to engage us that we become human. No one is a separated and independent entity—this philosophical notion belongs to the same series as being and having. We become human with others. In a word, we are human *co-becomings*.

Japanese Zen master Dōgen (1200–1253) discusses "taking an immediate reception here and now" in his earlier work *Gakudō-yōjinshū* (The Collection of Advices on Studying the Way) (c.1234). In this section, Dōgen states that there are two paths in Buddhist practices toward the enlightenment: "to visit masters and listen to their teachings" and "to make practices of sitting." The former path changes one's mind, while the latter path changes one's bodily experience. Two of them are sine qua non to complement Buddhist practices. In order to reach the state of "taking an immediate reception here and now," Dōgen proposes that we should contract our egos to open up a space for the others. In this space, we are immediately receiving the others including Buddha. The key word here is "others." It is obvious that in "visiting masters and listening to their teachings" Zen needs "masters" as others to guide us to be enlightened, although it is regarded as a symbol of self-powered Buddhism (Miyakawa 2013).

To illustrate this concept with a contemporary example, let us consider someone who is socially isolated, who rarely communicates with anyone. This person spends the entire day at home watching TV. We would say that this person just has limited capability. Capability is defined by Amartya Kumar Sen as "a person's actual ability to do the different things that she values doing" (Sen 2009).

So how can this person's capability be increased? For example, in a community which has no water supply, what would increase that community's capability more must be to teach the community how to dig a well rather than to give the community a drink vending machine. If that is the case, for the person socially isolated, which would increase one's capability more: buying him/her some DVDs to watch, or teaching him/her to ride a bicycle?

In the forthcoming society, the direction of our investment would be emphasized in the enhancement of human capability and the transformation of our way of living along with body and mind. Such investment will in turn provide fresh opportunities for us to change our habitus eventually. If, as part of the discourse on Society 5.0, we are to establish new indexes for better society, an urgent task is to find what could describe capability open to new chances for the way of living, i.e., habitus. For that sake, we cannot forget the dimension of engagement with others. To encourage engagement with others, it is important to foster an open attitude to receive others as Dōgen says, before letting socially isolated person to fall into self-consumed or self-destructed situation.

Once capabilities in a society are enriched, social mobility will increase accordingly. A rich society is often described in this way: the social mobility is much higher and the fixation of social class or social disparity is relatively weak. For example, Japan achieved a leap forward in social mobility during the Kansei era (1789–1801), when the government introduced a recruiting system based on the civil service examinations of Imperial China. The future society should have indexes for the fluidity of social mobility as well as the enrichment of capability. I would like to repeat again that it is important to pay much attention to the engagement with others.

Engaged Knowing

Having considered these points, we come to have an elaborated idea for the way of knowing in the forthcoming society.

In modernity, as symbolized in then university system, there occurred the first transformation of episteme, in which knowledge became systematized and proliferated across a nation-state. The characteristics of this modern episteme consisted of the historical investigations on origins and comparative studies based upon philology. It was this epistemic structure that interested Foucault in his economic analysis of labor, his biological analysis of life, and philological analysis of language.

After entering the twentieth century, the second transformation of episteme occurred. It was the result of society shifting its capitalistic object from things to events. The difference as digitalized information became important in this new episteme. University system also changed to reflect this transformation. The main procedure in university is now based upon information processing in the realm of engineering. Meanwhile, the humanities and arts which once guided the modern episteme are on the decline.

However, such a contemporary episteme has once again reached a turning point today. As mentioned at the beginning of this chapter, when Society 5.0 is realized in a future with its advanced technology that far exceeds human abilities, our contemporary episteme would be taken away from us.

If that is the case, we will confront the third transformation of episteme, in which "engaged knowing" could be introduced to enhance human *co-becoming*.

Thomas P. Kasulis has identified the formidable potential of engaged knowing in Japanese philosophy. This is what he says about Kūkai in his book entitled *Engaging Japanese Philosophy: A Short History*:

> Kūkai's intention was instead to know reality somewhat like how we *know a person*. Not to be confused with knowing *about* a person (which derives from reading and hearing about that individual), truly knowing a person involves some shared intimacy. To know another is to be inside that person's world, to interact or overlap with the person in such a way that the other person becomes part of your own life. Rather than objectifying the other, you share something with the other.
>
> Even in knowing an object, there can be a difference between a detached and engaged form of knowing. For example, skilled craftspeople do not just know about their tools and their media; they know them intimately by working with them, modeling their technique after the exemplary masters of the craft. By that process, woodcarvers come to perceive the uniqueness of each piece of wood and each chisel. They work *with* the wood based on an engaged, embodied knowing that allows the wood, the chisels, the artist's hands, and the artist's mind to be a harmonious whole, a single act of engagement.
>
> Similarly, when Kūkai left the academies on his quest to understand, he wanted to engage the world intimately, not as a detached observer. He wanted to know all of reality the way a potter, not a geologist, knows clay. By the time he returned from China, Kūkai had experienced firsthand the difference between the two kinds of knowing and was ready to explain it as the contrast between *exoteric* and *esoteric*. (Kasulis, Thomas P., *Engaging Japanese Philosophy: A Short History* [日本哲学小史], Honolulu: University of Hawaii Press, 2018, pp.108–109)

Kūkai wanted to know everything. To him, "detached knowing" was not enough. Instead, he advocated "engaged knowing." It is an intimate knowing in which we share our secret with close friends. According to him, this is what esoteric Buddhism is all about.

It is important for us to live as if we were Kūkai. To this end, it might help somehow to synthesize his teachings in a philological way or it might be interesting to design Kūkai-like AI robots, who could teach us on esoteric Buddhism in a way relevant today. However, those approaches are just "detached knowing" in which we are still spectators to our world. Along with Kūkai, we must train ourselves to engage with the other, understanding that, as Kūkai said, "the other person becomes part of your own life." This task is indeed a capital matter to us.

The Human Co-becoming

As a conclusion, I would like to summarize my argument in this chapter. In order to ensure that Society 5.0 does not become a dystopian society, we have to redefine the modern concept of humanity and find a path toward the human co-becoming with others. Nonetheless, this path is not so easy, because humans are open to possibilities to transform themselves into any direction including undesirable one. In other words, we do not have a fixed *telos* for co-becoming.

Fortunately, however, we have plenty of precedents to guide us in this way of human co-becoming. Of these, I intentionally pick out some Japanese cases such as Dōgen and Kūkai, as they offer insights about human co-becoming. To be sure, there are countless other examples throughout the world. Dōgen and Kūkai themselves both spent time in China, which to these Japanese visionaries represented a major "other," and this experience might have spurred them on in their pursuit of "engaged knowing." As long as human co-becoming is connected with our capability and social mobility, it will be much more enriched through the attitude to embrace plural and different languages and worldviews.

It would be wonderful indeed if our ancient knowledge like that of Dōgen and Kūkai, which is far prior to the epistemes of the nineteenth and twentieth centuries, turns up again in the future society in a new form.

References

Akerlof GA (1978) The market for 'lemons': quality uncertainty and the market mechanism. In: Uncertainty in economics. Academic Press, New York, pp 235–251. This issue was discussed by George Akerlof in "the market for 'lemons'" ("lemon" refers to a low-quality car)

Ames RT (2010) Confucian role ethics: a vocabulary. Chinese University Press, Hong Kong

Bresnahan TF, Trajtenberg M (1995) General purpose technologies 'Engines of growth'? J Econ 65(1):83–108

Cabinet Office (2016a) See Note 2 on page 13 of the 5th Science and Technology Basic Plan (Reference (Cabinet Office 2016a) in "Introduction")

Cabinet Office (2016b) See page 14 of the 5th Science and Technology Basic Plan (Reference (Cabinet Office 2016a) in "Introduction")

Keynes JM (2010) Economic possibilities for our grandchildren. In: Keynes JM (ed) Essays in persuasion. Palgrave Macmillan, London, UK, pp 321–332

Masuda Y (1989) Kikai kaihatsusha (Opportunity creators). TBS-Britannica, Tokyo

Miyakawa K (2013) Satori/kotoba/shūgyō (Enlightenment, language, training), Kodansha, Hon, August 2013 issue; ibid (2013). 'Konshin' to wananika (What is *konshin*?), Kodansha, Hon, September 2013 issue

Ohashi H (2017) Fudōsanryūtsūgyō to sangyōsoshiki: Kongōnimukete no kenkyū memo (Real estate transactions and industrial organization: research notes for the future). In: The Land Institute of Japan (ed) Kizonjūtakushijō no kasseika (Enlivening the existing housing market). Toyo Keizai, Tokyo. In a previous work, I discussed the same issue in the context of real estate:

Ohashi H (2018) Seisanseikōjō to aratanafukakachi no sōshutsunimukete no shiten (Perspectives for enhancing productivity and creating fresh added value], Policy Research Institute, Ministry of Finance, Inobeishon o tsujita seisanseikōjō ni kansuru kenkyū kai (Seminar on enhancing productivity through innovation)

Rifkin J (2014) The zero marginal cost society: the Internet of Things, the collaborative commons, and the eclipse of capitalism. St. Martin's Press, New York

Rosement H Jr, Ames RT (2016) Confucian role ethics: a moral vision for 21st century? National Taiwan University Press, Taipei

Sedlacek T (2013) Economics of good and evil: the quest for economic meaning from Gilgamesh to Wall Street. Oxford University Press, New York

Sen AK (2009) The idea of justice. The Belknap Press of Harvard University Press, Cambridge, Mass.

The regulation on promoting fairness and transparency (2019). An example is the regulation on promoting fairness and transparency for business users of online intermediation services. https://ec.europa.eu/digital-single-market/en/news/regulation-promoting-fairness-and-trans-parency-business-users-online-intermediation-services. Accessed 5 Jun 2019

Toffler A (1984) Future shock. Bantam, New York

Chapter 7
Interview: Creating Knowledge Collaboratively to Forge a Richer Society Tomorrow—An Innovation Ecosystem to Spearhead Social Transformation

Abstract As social problems at home and abroad grow increasingly complex and diverse, the Government of Japan is pursuing its vision of Society 5.0, the supersmart society that balances economic advancement with the resolution of social problems and where all can live comfortable lives. Meanwhile, the UN has advocated Sustainable Development Goals (SDGs) to address global challenges and has called upon industry to contribute to SDGs through business activities.

How can the R&D efforts of universities and businesses spark innovation and accelerate the pace of social transformation? This question was discussed in the following dialog between the University of Tokyo's Makoto Gonokami and Hitachi's Hiroaki Nakanishi, both members of the Growth Strategy Council.

Keywords Public–private–academia collaboration · Innovation ecosystem · IoT-led digital revolution · SDGs · Social transformation

As social problems at home and abroad grow increasingly complex and diverse, the Government of Japan is pursuing its vision of Society 5.0, the supersmart society that balances economic advancement with the resolution of social problems and where all can live comfortable lives. Meanwhile, the UN has advocated Sustainable Development Goals (SDGs) to address global challenges and has called upon industry to contribute to SDGs through business activities.

How can the R&D efforts of universities and businesses spark innovation and accelerate the pace of social transformation? This question was discussed in the following dialog between the University of Tokyo's Makoto Gonokami and Hitachi's Hiroaki Nakanishi, both members of the Growth Strategy Council.

The original version of this chapter was revised: This book was inadvertently published with the incorrect license type CC BY 4.0 and the Open Access License has been amended throughout the book to the correct license type CC-BY-NC-ND. The correction to this chapter is available at https://doi.org/10.1007/978-981-15-2989-4_9

Society 5.0, https://doi.org/10.1007/978-981-15-2989-4_7

Makoto Gonokami, President of the University of Tokyo (left). Hiroaki Nakanishi, Chairman of Hitachi (right).

7.1 Society 5.0 Is About a Common Goal

Nakanishi Society today faces a myriad of problems, and there is a growing impetus for the social reforms necessary to address these problems. This situation was reflected in the 5th Science and Technology Basic Plan, approved by the Cabinet on January 22, 2016. Under the major theme of "future industry and social transformation," the Plan outlined a firm commitment to "Society 5.0," a vision that calls for more effective R&D so as to stimulate disruptive innovation, and that calls on Japan to lead the world in implementing a supersmart society. You and I were involved in drafting this plan as members of the Council for Science, Technology, and Innovation. May I ask you first of all just to recap how you define Society 5.0 and what you see as the context behind it?

Gonokami The time when we drafted the 5th Science and Technology Basic Plan was a time of mounting expectations for an IoT-led digital revolution. After reflecting on the outcomes of the 4th Basic Plan, we decided that "Society 5.0" should be a key term to encapsulate our future vision, in that it expresses the idea of taking society as a whole to a new place.

You and I then worked to flesh out ideas on Society 5.0 as members of the Growth Strategy Council, which had been established in the Headquarters for Japan's Economic Revitalization in September 2016. We soon realized that a digital revolution would involve a disruptive transformation of all industrial and social systems. This phenomenon could be called super-smartification—a situation where the use of Big Data and other new value-generating processes lead to seismic changes in the very fabric of society.

We recognized that we should not sit back and watch as technology reshapes society. Instead, we should actively seize the opportunity and lead the process. Social transformation is an urgent task. This is particularly true in the case of the shrinking and aging population, a problem that we must address within the space of a few years. We saw this situation as an opportunity for action. We knew that we could only overcome the problem through a game-changing solution. To this end, we would need to develop necessary technologies and services ahead of the rest of the world and highlight the tasks to tackle. We thus took stock of Japan's existing social values and its strengths and then discussed the implications of social transformation—what shape will society be in following the transformation, and what should we be doing now? We decided that the term "Society 5.0" would help focus minds in this direction. It was thanks in large part to you that the term caught on.

Nakanishi You give me too much credit. For my part, I understood how important the concept was thanks to my experience in Hitachi. As part of a group-wide reform project, I had introduced a social innovation business, a business that focuses on taking society in a new direction. Initially, many people in my company were skeptical, saying that they were unsure what social innovation was all about. Even so, I made a point of using the term, as I believed we needed a concept to indicate an overall direction and vision—something you cannot do if you only ever talk about specific technologies.

The same goes for Society 5.0. The concept allows us to share a common goal, to create a new society together. We use the term "supersmart society" because we set our sights beyond a technologically driven society, to a more human-oriented society. In designing the architecture for such a society, we should not dictate from the outset what Society 5.0 should look like. As we develop the concept, we must give room for creative and innovative ideas to grow and allow new values to emerge.

Gonokami You just reminded me of something. Around 2 years before I became president of the University of Tokyo, I established the Innovative Center for Coherent Photon Technology (ICCPT) with funding from the Japan Science and Technology Agency's Center of Innovation. Coming from a background in laser physics, I wanted the ICCPT to spark a manufacturing revolution by integrating laser optics with materials technology. To that end, the ICCPT would work with other research institutes and with manufacturers involved in lasers and materials processing.

During Japan's high economic growth period in the second half of the twentieth century, Japanese manufacturers succeeded in producing high-quality goods at low cost through a combination of automation and quality control. However, the proliferation of standardized, mass-produced goods has resulted in a society where people's lives are shaped by material goods. The task now is to innovate manufacturing technologies so that we shift to a society where material goods are shaped by people's lives, where manufacturers produce high-quality customized goods as cheaply as mass-produced goods.

I believe that the digital revolution can create a sustainable society, one that uses resources efficiently according to the specific needs of individuals. The shift toward customization in manufacturing is, along with other developments such as tailor-made medicine and flexibly working practices, part of a broader shift from product-focused thinking to individual needs-focused thinking, and it is this broad shift that holds the key to social transformation. The digital revolution is not simply a matter of the tools we use; it is something that will radically transform the structure of society itself.

7.2   Fostering the Mind-Set to Try Something New

Nakanishi To create a digital revolution that will qualitatively transform society and create new values, we will need a new human resource development strategy.

Gonokami As I said before, we face the urgent issue of a shrinking and aging population. If we are to find a game-changing solution to this issue in the space of a few years, we need to have more minds working on it. To this end, universities need to change. It is no longer good enough just to launch young people into the workforce. Universities should invite workers to return to academia and collaborate with academics on solutions to challenges. I do not mean that more people need to return to university to restart their education. I am saying that universities should actively encourage a form of recurrent education in which adults in the workforce join forces with academics to come up with ideas on problems.

Some educators argue that children must learn programming and foreign languages from elementary school in order to prepare themselves for future changes, but this is a little unreasonable for the youth. Instead of telling the next generation, whose numbers are small, to support our generation in the years ahead, we should lead by example and take the necessary actions ourselves. We need to foster a mind-set in which, instead of fearing change, one is willing to try something new and not be scared of trying something that no one else is doing.

It is people who generate new values. Therefore, universities, insofar as they educate people, have an ever-greater role to play. Universities should play a central role in driving a paradigm shift in collaboration with different sectors of society.

Nakanishi Academia has a vital role to play in reforming attitudes toward globalization. Businesses must globalize, but so too must our culture and our daily life itself. There must be a qualitative shift in our relationships with the wider world. If we engage and communicate with people of different ethnicities and cultures, we can develop the robustness to adapt to an uncertain time and a dynamically changing world.

Gonokami The meaning of globalization is itself changing. Originally, the term suggested a homogenization process whereby developing nations adopt the models of developed nations. Nowadays, however, we understand globalization as an attempt to create a world where people from all walks of life can happily coexist. If we are to understand and respect people of different backgrounds, we must learn to see ourselves from a more proportionate perspective and in relation to others. That is why I encourage students to spend time studying overseas. For my part, I try to expand opportunities and support as many talented students as possible.

7.3 Innovation Comes from Melting Pot of Ideas

Nakanishi Structural change has begun in the industrial sector. The industrial boundaries are breaking down. If industries adopt global perspectives and think about things in global terms, they will be better able to plan ahead in anticipation of change. I am not saying that people in industry need to read a load of books; rather, they should learn by actually encountering and interacting with different people.

Gonokami Absolutely. People can access a vast array of information online, but that in itself will not be enough. In the years ahead, more and more value will be found in settings of flesh-and-blood encounters, such as university campuses. When people of different perspectives bring their particular experience and knowledge to the table and discuss face to face with each other, you get a melting pot of ideas from which innovation will emerge. We need to increase the opportunities for such interactions.

Nakanishi I could not agree more. We always seem to think alike! Japan's top universities also have amazing international potential. Is the University of Tokyo making efforts to globalize in some respects?

Gonokami The University of Tokyo exists to promote diversity in the world's knowledge. As such, we must clarify and communicate our values and role. If you compare the University of Tokyo to other universities around the world, you will notice that our university stands out in the fields of humanities and social sciences. This value is something that we are trying to communicate to a global audience. Integrating the humanities/social sciences, natural sciences, and technology is an essential step in creating a sustainable society that supports individuality. We are working to build new structures to facilitate interdisciplinary collaboration and transcend cross-epistemic boundaries with a view to producing globally unique value.

7.4 Industry–Academia–Government Collaboration for Building an Innovation Ecosystem

Nakanihsi Social and industrial structures are on the cusp of a paradigm shift not only in Japan but also around the globe. As you said at the start, to accelerate the transformation, efforts must be coordinated across sectors, and that includes academia, industry, and also the government. In June 2016, the University of Tokyo and Hitachi launched H-UTokyo Lab. to establish a shared vision and pioneer a new form of industry–academia collaboration. What do you think about this new innovation ecosystem?

Gonokami One of the most important tasks of my presidency is to enhance the foundations for industry–academia collaboration. My own research lab has so far sent over 100 students into the workforce. Around 70% of the graduates went into the industrial sector. Judging from what they have told me since then, over the past 10 years, the industries have not fully utilized these graduates' abilities. As industry undergoes structural transformation, we need to ensure that human resources are employed in positions that suit their skill set and where they can realize their full potential, so that new value can emerge. Universities can help in this task as they understand their graduates very well.

Meanwhile, industry has ever-greater expectations of academia. Amid the toughening global competition and the need for quick results, businesses must keenly discern where to apply their strengths and what to invest in. Academics are good at taking a longer term view of things, a skill they gain in their fields of study. To deploy these long-term insights, we need a new form of industry–academia collaboration, one that takes into account the changes in the business environment. The first step is to establish the contractual structures that will enable businesses to invest in universities with peace of mind. We are already seeing results in this area, and so I look forward to seeing an innovation ecosystem develop in the years ahead.

Nakanihsi Yes indeed, the business climate is growing more complex, and in many cases, it is hard to see what the real issues are. Against this backdrop, businesses will not find solutions if they rely only on their own theories and hypothetical scenarios.

That is why there must be a broader ecosystem of ideas. The collaborative model that H-UTokyo Lab. advocates is one in which the top businesses and universities engage with each other, not only in technological projects, but also in sharing a common future vision and finding real-life applications for the university's diverse knowledge so as to forge shortcuts to solutions. Government should engage in the process too, as the future vision concerns social problems. Venture capitalists also have an essential role to play when it comes to financing. A paradigm shift and new industrial creation will be possible once the stage is set for these four actors to work together, with academic knowledge as the driving force. Having lagged behind its overseas counterparts in this respect, Japanese industry is now waking up to the need for such collaboration. So I feel we now have a great chance to promote a new homegrown model of an innovation ecosystem underpinned by industry–academia–government collaboration.

Gonokami As you said, venture capital has a vital role to play. In 2004, our university founded a venture capital firm called the University of Tokyo Edge Capital (UTEC). UTEC has supported the commercial application of research findings and worked to build up a body of knowledge necessary for such. Although we still have some way to go to catch up with the top US universities, we have helped form around 300 startups, 17 of which are now listed companies. The aggregate market value is 1.4 trillion yen. I anticipate that industrial sector will increasingly support the commercialization of research findings through carve-out startups, so I am sure that industry–academia collaboration will be possible by sharing our expertise in launching startups.

Nakanishi In 2015, you released "The University of Tokyo: Vision 2020." The basic principle underlying this vision is "Synergy Between Excellence and Diversity: Acting as a Global Base for Knowledge Collaboration." Would you say that this principle is the same idea as the innovation ecosystem?

Gonokami Having led East Asia in terms of academic and industrial innovation, Japan is an ideal place for creating new knowledge with cross-border value, and I want the University of Tokyo to be a base for such knowledge creation. In an ecosystem that produces knowledge with direct economic value, Japan will always have a role to play.

Nakanishi I want H-UTokyo Lab. to bring industry–academia collaboration into a new phase and promote an ecosystem that engenders innovation. Energy is one of the areas in which we are pursuing research. This field involves numerous stakeholders and the goal of the research is not to benefit Hitachi alone. In this respect, I believe we can form an ecosystem core.

Gonokami The need to find a game-changing solution is a matter of urgency, so we must press ahead with the task of finding some practical application for our research findings.

7.5 Linking Research Activities to SDGs

Nakanihsi So far, we have discussed Society 5.0 as a national vision, but as industry, academia, and government work to produce new values, I believe they should be guided by the worldwide future vision contained in the UN's SDGs. You wasted no time in incorporating the SDGs into the university's business strategy. What do you see as the role of leaders in contributing to the SDGs?

Gonokami Earlier, I mentioned our Vision 2020, which outlines the aim of creating synergy between excellence and diversity. Under this aim, we have sought to produce excellence from diverse activity. In other words, we have encouraged researchers and students to act based on their free ideas, believing that a critical mass of such free agency will provide a driving force to move society toward a better direction. We understood that we can only achieve this goal if the researchers and students are committed to a common vision at higher levels. It just so happened that the UN announced its SDGs around this time (in 2015), and so we decided to set the SDGs as our goals. When we announced these SDG-inspired goals, it made little impact at first, but attention picked up as more people realized how important SDGs are in attracting active investment and enlivening the economy.

You were part of this trend too; didn't Keidanren update Charter for Good Corporate Behavior and Implementation Guidance to incorporate the SDGs? Coupled with the global rise in ESG funding, SDGs are encouraging businesses to step up their efforts in promoting sustainable corporate value.

At the University of Tokyo, we started with getting the teaching staff to record which of the 17 SDGs correspond to their teaching or research activities so that we could map out these activities. Over 150 activities were recorded. Our activity map helped us visualize the areas where our university is doing well. It also allowed to identify interlinked research projects, where we could encourage interdisciplinary research.

SDGs are indispensable in that they drive economic activity while also fostering a more accommodating form of capitalist economic development. It is in this that their value lies. Universities have a role to play in adeptly matching up SDGs with research activities so as to derive solutions to the challenges.

Nakanishi This overlaps with what you just said, but I think the reason SDGs have gained solid traction is because more people are adopting global perspectives. Businesses have traditionally had the notion that as they earn profits by imposing a burden on the environment, they can make an environmental contribution in return. However, as the idea that the whole world is connected has permeated deep into society, this traditional approach is not producing value anymore. Businesses need to go back to basics in a sense. They need to share common goals and plan environmentally sound business activities with global perspectives from the outset so as to achieve sustainability in the true sense.

Thanks to SDGs, business leaders are increasingly realizing that challenges related to the environment and energy are related at a fundamental level to the problem of poverty. In this respect, the role of leaders in the industrial sector is to restructure business.

Gonokami A notable feature of university research is the diverse range of timescales. Some research projects are very short-term, but there are also some projects that a university can sustain over a 100-year or even a 200-year time span. Industry, too, once operated with long-term perspectives, but economic cycles have become shorter, making it hard for businesses to sustain long-term research and business projects. SDGs may help reverse this trend. If industry and academia back each other up, they may create an environment in which businesses can pursue projects that have a greater range of timescales, including longer term projects, with peace of mind while retaining economic rationality.

Nakanihsi So you are saying that long-term projects should not just be left to the universities but conducted jointly.

This year (2018), Hitachi's R&D organization celebrated its centenary. What role do you expect corporate R&D teams to play in the new system of industry–academia collaboration?

Gonokami Such milestones are great opportunities to take stock, and review how things are being done.

In 2017, we celebrated our 140th anniversary as a university. We looked back over our history, segmenting it into two 70-year periods, and named the ensuing third 70-year period "UTokyo 3.0." We decided that in UTokyo 3.0, we should be a university that spearheads a transformation toward a better society, a society in which individual free agency underpins the stable development of humankind as a whole.

The academic and business communities each have their own role to play. I hope that business leaders for their part will make continued efforts to pursue long-term research as part of their corporate activities. Under the vision of Society 5.0, let us work together to achieve an industrial and social transformation that attributes value not just to technology but also to the wisdom that truly serves humanity.

Nakanishi Yes, let us co-create knowledge and thereby spearhead the transformation. In 2015, we restructured our R&D organization to make it more customer oriented. Our task at Hitachi now is to prepare our vision for the next 100 years, and you have given some valuable hints in this regard. Thank you very much for your time today.

President of the University of Tokyo
Makoto Gonokami, D.Sc.

1982: Earned an M.Sc. in Physics at the Department of Physics, Faculty of Science, The University of Tokyo
1983: Research Associate, Department of Physics, Faculty of Science, The University of Tokyo
1990: Associate Professor, Faculty of Engineering, The University of Tokyo
1998: Professor, Department of Applied Physics, Faculty and Graduate School of Engineering, The University of Tokyo
2010: Professor, Department of Physics, Faculty of Science and Graduate School of Science, The University of Tokyo
2012: Vice President of the University of Tokyo
2014: Dean of the Faculty of Science and Graduate School of Science, The University of Tokyo
2015: President of the University of Tokyo

Chairman and Representative Executive Officer of Hitachi, Ltd.
Hiroaki Nakanishi

1970: Joined Hitachi, Ltd.

2003: General Manager of Global Business, Vice President, and Executive Officer, Hitachi, Ltd.

2004: Senior Vice President and Executive Officer, Hitachi, Ltd.

2005: Chairman and CEO, Hitachi Global Storage Technologies Inc.

2006: Executive Vice President and Executive Officer, Hitachi Ltd.

2010: Representative Executive Officer and President, Hitachi Ltd.

2014: Representative Executive Officer, Chairman and CEO, and Director, Hitachi Ltd.

2016: Executive Chairman and Representative Executive Officer, Hitachi Ltd.

∗The above is an extract from "Innovators," a special edition (March 2018) of *Hitachi Review* to commemorate the centenary of Hitachi's R&D Division.

Chapter 8
Issues and Outlook

Atsushi Deguchi and Kaori Karasawa

Abstract As a final part, this chapter discusses the goals and the issues in the process of realizing Society 5.0 from the view of happiness of human being in harmonizing with the society, and concludes by overviewing the significance of Society 5.0 and its outlook as a policy for the data-driven society promoted by digital revolution.

Section 8.1 mentions the issues in the happiness to be provided with human being through data-driven society, and points out that it is needed to clarify the approach through which each person will be able to obtain his/her own happiness by approving the data-driven technology implementation and harmonizing with the data-driven society. In addition, it mentions the issues in the coexistence of the free choice by persons and the social control, and suggests that we should apprehend the moral questions to be considered in the process of realization of the data-driven society.

Section 8.2 summarizes the social meanings and significance of Society 5.0 as a vision originating in Japan to be aimed with the implementation of advanced digital technology beyond the conventional smart city ideas. Consequently, it concludes by emphasizing on the importance of sharing the concept of "people-centric" in order to realize both the social problem solution and the economic growth as mentioned in the original definition of Society 5.0 in the Comprehensive Strategy on Science, Technology and Innovation for 2017.

Keywords Citizen-based innovation · Happiness · Regional revitalization · Technology-led social vision · Well-being

The original version of this chapter was revised: This book was inadvertently published with the incorrect license type CC BY 4.0 and the Open Access License has been amended throughout the book to the correct license type CC-BY-NC-ND. The correction to this chapter is available at https://doi.org/10.1007/978-981-15-2989-4_9

A. Deguchi (✉)
Department of Socio-Cultural Environmental Studies, Graduate School of Frontier Sciences, The University of Tokyo, Tokyo, Japan
e-mail: deguchi@edu.k.u-tokyo.ac.jp

K. Karasawa
Division of Socio-Cultural Studies, Graduate School of Humanities and Sociology, The University of Tokyo, Tokyo, Japan
e-mail: karasawa@l.u-tokyo.ac.jp

8.1 Question of Happiness: Harmonizing Individual and Societal Interests

Humans and Happiness in Society 5.0

A supersmart society is where cyberspace is merged with the physical space (real world). That is what Society 5.0 is supposed to be. Underpinned by AI and Big Data, society will transform radically. The society to aim for is one that addresses the deep-seated hindrances to sustainability so that the people can lead a fulfilling and happy life. One of the keys to achieving this vision is to find how to create the right environment for society's inhabitants. This task requires planners to discuss the direction of new urban environments, consider how to design the society, collaborate with academics from different fields of study, and integrate cutting-edge technology and analytical approaches that are related to manufacturing and community development. Once the planners start creating an environment that supports a better life and establishing the institutional groundwork for building a sustainable society, they will have made an important step toward Society 5.0.

Where does the human individual fit into all this?

The literature on Society 5.0 is replete with references to humanity, people, and individuals. For example, there are frequent expressions such as "enhanced humanity," "respect for human dignity," a "human-centered society," "people-friendly," "greater freedom for the individual," and a society customized to "diverse human preferences." This language suggests that individuals' happiness is pivotal in designing environments and institutions, and that Society 5.0 must be designed in such a way to attain this objective.

Free and effective use of information, coupled with innovation in environmental and institutional designs, will emancipate individuals from the restrictions that hinder them from living a better life. Once freed from these restrictions, individuals can fulfill their desires and needs without undermining the sustainable development of society as a whole. Such a society is a happy society, for the individuals therein gain the mental health that comes with a satisfying and meaningful everyday existence, in addition to physical health. Such is the people-centric vision that Society 5.0 advocates.

The Challenge of Reconciling Individual and Societal Interests

So we share and commit to this marvelous future vision, but what must be done to make it a reality?

When we take a dispassionate view of the situation, we will see a host of challenges that today's society must overcome, including the depletion of energy resources, environmental degradation, elderly care needs, and shrinking workforce. These problems would not naturally disappear just because society elevates from 4.0 to 5.0. They will only get worse unless we find effective solutions. It is more urgent than ever to construct a society that sustainably reconciles the outcomes of individuals' behavior with the common good.

However, it is no easy task to keep in mind the common good and strike the right balance between empowering and controlling individuals' choices. Humans are autonomous agents who exercise free choice, and these free choices cannot and should not be curtailed lightly. Yet this problem is exactly why we need a serious discussion on how to reconcile individual and societal interests. Such a discussion will be a critical step in defining what happiness means in Society 5.0.

Defining Happiness

We return to the question of what happiness is, a question that has occupied the minds of thinkers the world over since time immemorial. In designing a society, we must work out the conditions underlying happiness or well-being. What do we need to be happy? Let us examine the outcomes of the discussion process organized as part of a national project, as these outcomes constitute, to a certain extent, consolidated findings on the matter.

In 2010, the Cabinet Office gathered social psychologists, economists, and other experts and launched the Commission on Measuring Well-being. The commission reviewed the literature from Japan and overseas and selected certain metrics for measuring well-being. It then released its report in December 2011. The report is available online (Cabinet Office 2011).

According to the report (those interested in the finer details can read the full report), although it may vary with factors such as age, there are three common requisites for subjective well-being, each of which is predicated on communal sustainability. The first is socioeconomic condition, which includes wealth, income, work, housing, education, security, and safety. The second is health, which includes physical and mental health. The third is relatedness, which includes bonds with family, bonds with community, and lifestyle.

A Happier Society

These three requisites will likely remain the same whatever the times are. Given this, if advances in AI and digital transformation lead to urban environments, which are more resident- and worker-friendly, or if advances in healthcare allow us to live healthier lives, society as a whole will be much happier. In other words, if, as we move toward Society 5.0, we manage to improve socioeconomic conditions and promote better physical and mental health, we can achieve a happier society at least in these two aspects.

But this change will entail the dilemma I mentioned earlier: how to reconcile individual and societal interests. Even if technology streamlines our systems and makes us richer, we will still face the same challenge of having to distribute limited resources. Although it might become easier for individuals to seek comfort and fulfill their specific needs, such a society will not be sustainable if it gives individuals free reign to rampantly pursue consumption, unfairly monopolize resources in pursuit of their own happiness, or otherwise exploit its systems. Aside from individuals' pursuit of happiness, there must be a moral code directed toward the common good, and individuals must act in accordance with it. Technology and data alone are not enough to ensure that Society 5.0 is a happy society. Social design must emphasize the task of harmonizing freedom of behavior with behavioral regulation.

Behavioral regulation must accord with human nature; otherwise, there will be no true harmony between individual and society. Respect for human dignity would be undermined if society restrains individual freedom, or if individuals defraud society. A happier society is achievable when individuals, in their pursuit of happiness, exercise freedom of choice according to their own values, and when these individuals' behaviors are tempered by a moral code that enjoins sustainability.

Social Design and Relatedness

Social design must also consider the third condition of well-being, relatedness. Relatedness is significantly shaped by things like housing, workplace, digital environments, human services, and AI substitution.

But, how, in the first place, will the supersmart society, one formed by "merging the cyberspace with the physical space (real world)," change human relationships and what new communities will it create? Will we become more homogenous in terms of class and values, or will we become more diverse? We might use the extra free time to socialize with those who are close to us, but in doing so, we might grow more distant from those less dear to us. Will we spend more time engaging in flesh-and-blood social interactions, or will we interact more with AI friends? In asking these questions, one realizes how little we currently know about how Society 5.0 would influence relatedness.

The opaqueness of this issue makes the task of social design all the more important: the task of framing Society 5.0 as a society with a relatedness conducive to well-being. But once again, the harmony between individual and society will be jeopardized if all priority is placed upon allowing individuals to pursue, in the here and now, the kinds of relatedness that they believe will yield comfort. Suppose, for example, that everyone interacts only with people who share their values and avoids everyone else. Such a scenario may be mentally comfortable. However, it would also create exclusive cliques, giving rise to inequality and discrimination. It would also deter tolerance and the creativity that arises from diversity. Consequently, human life may become poorer.

An important thing to remember is that social design often affects relatedness in unintended ways; there are side effects. Planners might design a smart city to be convenient, safe, and comfortable, and then present it to people, but would people move to this city randomly? Or would the planners ultimately, if unwittingly, select for certain groups, such as members of certain socioeconomic statuses or holders of certain values?

Unintended consequences are difficult to predict. They are especially difficult to predict when it comes to human relations, owing to the myriad of social phenomena interweaving such relations. All the more reason, then, to be extra mindful of how society is the aggregation of individual relations and of how happiness in Society 5.0 must be grounded in the harmonization of the two.

Free Choice and Social Regulation

For Society 5.0 to be a richer, more comfortable society, we might start seeing the pursuit of greater comfort and wealth as something that is perfectly normal. When people have greater freedom to choose the things they like, perhaps they will make more selfish choices. Thus, in order to reconcile individual and societal interests in a way that achieves greater happiness, we must at some point regulate individuals' behavior.

A society that proclaims a high level of happiness (in terms of comfort, convenience, wealth, and health) for the many is a society that unleashes people's desires, such that people have a much higher level of demand or a stronger desire to freely act to get the things they value. However, when people are freer to pursue the things they want, they will sometimes harm the common good, so it is necessary to control individuals' behavior to some extent. In the second half of this section, I explore how society can regulate the behavior in a way that accords with human nature.

The Pitfall of Rewards and Punishments

One rather crude way of controlling behavior is to offer rewards and punishments. Undeniably, rewarding certain behavioral outcomes with money or nonmonetary compensation can powerfully shape behavior. Even if some claim that they do not work for money, which is one side of truth about human kind, money is a central fact of human life. No society could exist without a system of rewards and punishments, and we live our lives within such a system.

Both rewards and punishment entail certain social costs. These are the costs associated with delivering the rewards and punishments and that of monitoring whether they are being delivered appropriately. These costs could be reduced by incorporating technology into institutional environment. In this way, Society 5.0 could condition human behavior with a system of rewards and punishments that is more efficient than its previous iterations.

But herein lies a pitfall. Relying on a system of rewards and punishments may undermine the goal of a people-centric society. That is, it may run counter to respect for behavioral autonomy and the will to seek freedom—the idea that people should act in a way that is true to their inclinations and values. Once people start believing that their behavior is being conditioned by a reward or punishment, they may lose their intrinsic motivation and start exhibiting reactance (more on this later).

Intrinsic Motivation

There are two main types of motivation behind human behavior: intrinsic and extrinsic motivation. Intrinsic motivation is the desire to act based upon one's interests, inclinations, or values. Extrinsic motivation is motivation that comes from outside the person, such as from rewards, punishments, or coercion.

Too much extrinsic motivation can kill off intrinsic motivation. For example, suppose that a group of people desire to save energy. If these people live under a system that rewards energy-saving efforts, they will naturally make an effort to save energy. However, once they start attributing their efforts to the reward society offers them, they will cease to believe that they are making energy-saving efforts because they intrinsically want to make such efforts. Consequently, their intrinsic desire to save energy is undermined. Likewise, when you start believing that you are working for the pay, you will in many cases lose interest in the work itself.

Humans always seek a reason for why they are doing something, and when a reason becomes prominently apparent, other possible reasons get pushed to the wayside. Thus, once individuals start seeing extrinsic inducements, such as rewards and punishments, as the basis for their behavior, they will get the notion that they are not acting this way as a result of their inclinations or values.

Reactance

We believe that we have a freedom to choose our own action. As such, we react defiantly when it seems that someone is taking away our behavioral choices or forcing us to choose a certain action.

This is called reactance. One problem with reactance is that the inability to exercise a certain option can make that option appear more attractive than it should be. Something that a person would have chosen, had they been able to act freely, will start to appear all the more attractive as a result of the person having been potentially able to obtain it. Another issue is that reactance creates a mounting desire to restore one's subjective freedom of choice. A person may naturally incline toward certain desirable behavior, but if you attempt to induce that person to perform this behavior, they might make an alternative choice despite their original preference for the desired behavior. To relate this phenomenon to the example of energy saving, when people believe that they are being induced into saving energy by a system of rewards and punishments, they will start seeing wasteful energy use as all the more attractive, and may deliberately waste energy when no one is watching.

Design That Fosters Desirable Inclinations and Values

Intrinsic motivation and reactance may unleash the inner devil. But they are essentially linked to that most essential part of humanity: our autonomy and freedom of choice. They are key to honoring the dignity of humans as autonomous agents who act according to their perception, beliefs, and values. To be people-centric, society must have its environment and institutions designed upon the premise of human autonomy.

Therefore, planners must always be circumspect about the extent to which they rely on rewards and punishments. To ensure the sustainability of the system, the utmost care must be paid to the question of how much you circumscribe human behavior. If looser regulation is possible, rewards and punishments should not be introduced rampantly. Instead, there should be a more gradual system of inducements (nudges are an example of this) to promote behavior that leads to a harmonization of individual and societal interests. This strategy will help ensure that people behave in a way that is true and natural to themselves.

This strategy is advantageous because it fosters the inclinations and values that align with the desired behavior. I claimed earlier that if people believe that their behavior is motivated by a reward, they will be unlikely to believe that they are motivated by their own inclinations or values. However, the reverse is also true; when people do something without any rewards, they will attribute the cause of their behavior to their inclinations and values.

Thus, if society has subtle inducements (as opposed to rewards and punishments) under which individuals choose to act in the desired way, these individuals will recognize that their inclinations and values naturally align with the desired action. This strategy may therefore succeed in conditioning individuals' behavior without undermining the human desire for autonomy and free choice. At the same time, in empowering people to act in accordance with their inclinations and values, the strategy may also help ensure that individuals' behavior aligns with the interests of society as a whole.

Of course, this may not be so easy to accomplish in practice. If we could quickly and easily foster the inclination and values underlying the desired behavior, we could solve many of the social problems before they become too serious. We know from common sense as well as from research into human behavior that people will frequently act contrary to how you want them to. Therefore, I neither propose a simple recipe for behavioral regulation nor am I saying that rewards and punishments should never be used.

The key point is this. All societies need to control individuals' behavior in some way, and Society 5.0 likewise must do so, deploying all available wisdom to this end. But Society 5.0 must do so in a manner that accords with human nature—not just to prevent unexpected misfires, but to forge the way to a happier society, where there is harmony between individual and group interests. Planners must adopt such a perspective when designing the environment and institutions, as the principle of honoring human dignity requires no less.

Finally, Some Outstanding Moral Questions to Consider

We have discussed how we can achieve happiness and well-being in Society 5.0, and how to this end we must harmonize individual and societal interests in a manner that accords with human nature. Finally, I want to raise some moral questions. What underlying norms and principles should a society refer to when deciding how to guide its members' behavior? What kinds of behavior should we allow society to regulate? Who has the right to subtly induce behavior in others?

The world is already awash with inducements, including online ads. Against this backdrop, it may be desirable for benevolent planners to induce behavior, taking into account the common good as well as commonplace value judgments. However, some may intuitively feel aversion or dread toward a society that uses a system of ploys to make people behave in a certain way without them even suspecting that they are being conditioned.

There are no clear answers to the above issues. The absence of answers should not be an excuse to ignore the questions or to shelve all the work we must do to harmonize the interests of individuals and society. Although the answers may elude us, we must keep seeking them out.

We must do so because we have a duty to the future generation who will live in Society 5.0. Those who introduce new technology or design institutional arrangements and those who debate the shape of Society 5.0 must consider, from various perspectives, which of the available options would be more judicious or appropriate, if not absolutely ideal. Such an approach will help ensure that the future society honors its members and delivers to them happiness and well-being.

8.2   Significance of Society 5.0 and Its Outlook

Up to now, we have discussed the concept and nomenclature of Society 5.0. We have also discussed the basic approach to making Society 5.0 a reality, the basic approach to technological development, and how we might achieve a people-centric society. In this section, I outline the social significance of Society 5.0 as well as the outlook and challenges.

Vision for a Society Driven by Technology

As outlined in the Science and Technology Basic Plan, Society 5.0 simply presents a vision of a society driven by science and technology. A supersmart society where cyberspace is merged with the physical space (real world) is underpinned by technology for gathering and collating data within a cyberspace architecture, and by technology for converting the data into knowledge and reintegrating it into the physical space (real world). This book has focused on such a technology and introduced information integration architecture (Chap. 4) and approaches for transforming urban habitats (Chap. 5).

This technology targets data collected from the physical space (real world). With this technology, all kinds of data, including that related to energy, transport, shopping history, emissions, and other facets of urban environments, get stored in cyberspace. In its raw state, the data is just a series of digits. However, the technology processes the data into meaningful information and then into knowledge. This knowledge then actively influences the physical space (real world). In this respect, the supersmart society, one formed by "merging the cyberspace with the physical space (real world)," is essentially a more advanced form of the knowledge-intensive society and data-driven society.

The difference is that the future society in which the technology will be used is a people-centric society. Solutions to tackle social challenges (such as the super-aging society and the carbon-free society) may end up forcing people to make sacrifices. The technology in Society 5.0 is that which balances such solutions with the principle of a people-friendly society. Though the society is driven by science and technology, it remains people-centric. The researchers and engineers working in R&D must bear this point in mind: Society 5.0 is a vision of a science and technology-driven society, but the goal of this vision is a people-centric society.

Principle of People-Centric Society and How We Get There

What is a people-centric society? To recap, in the Government's Comprehensive Strategy on Science, Technology, and Innovation (STI) for 2017, Society 5.0 is described as a society that, "through the high degree of merging between cyberspace and physical space, will be able to balance economic advancement with the resolution of social problems by providing goods and services that granularly address manifold latent needs regardless of locale, age, sex, or language to ensure that all citizens can lead high-quality, lives full of comfort and vitality." This definition tells us two things.

First, it tells us that Society 5.0 is a sustainable society, one that balances the resolution of social problems (the interests of society as a whole) with people's need for security and comfort (interests of individuals). As the pressure mounts to deal with climate change, Japan now faces the urgent task of going even beyond the low-carbon society, to the zero-carbon society. As a developed nation with an aging population, Japan also faces an urgent task of coping with the super-aging society. Tackling these challenges without hindering people from living in security and comfort in the process is, for Japan, key to becoming a model of how to overcome the problems associated with a developed economy.

Chapter 2 discussed "Habitat Innovation," a framework for approaching this task. This framework helps steer policymakers away from solutions that force people to make sacrifices. It does so by breaking down the indices for solving social problems into three broad components (policymaking, technological innovation, and pursuit of QoL) and various metrics so as to highlight the optimum balance between what is best for society and what is best for the individual. The chapter then underscored the importance of industry–academia–government collaboration in each of the three components. Research on improving QoL has a particularly crucial role to play in promoting the people-centric society, and the humanities and social sciences can offer vital insights to shape our vision of society and humanity, an essential task in making Society 5.0 a reality.

The second thing this definition tells us is that Society 5.0 is an inclusive society, one that accommodates diversity and a multiplicity of preferences. Previous approaches have tended to emphasize economy and efficiency at the expense of capitalizing on the unique features of communities. When people live in homogenized residential environments where choices are limited, they may end up conforming to a cookie-cutter lifestyle. Amid diversifying preferences, Society 5.0 points to a society in which people have more freedom of choice in their residential environments and lifestyles, and are better able to enjoy their hobbies and leisure time. It is a society in which people access services that suit their specific preferences without segregating themselves from people of different preferences or of a different income level. Already we are using cash less and less, and we are shifting increasingly to nonmonetary and sharing economy, in which ownership of tangibles has less value. As society changes, individuals must too. As Chap. 6 argued, this society-level transformation challenges us to reevaluate our values and revisit the question of what makes us happy.

IT is driving change in systems related to the economy, education, and welfare, so another challenge is to devise new kinds of social structures. We must also have a deeper discussion on what makes individuals happy and how individuals and society should interface. The humanities and the social sciences have an important role to play in making Society 5.0 a reality, and once the discussion of these issues becomes open to the public, the Society 5.0 concept will start to permeate in the hearts and minds of the people.

Citizen-Based Innovation

Chapter 3 discussed the existing smart city concept, citing past cases where technologies such as smart grids have been applied in the energy sector. Traditional smart city models involve the practical application of techniques and technologies that use data in a particular sector (such as energy or transport). The supersmart society goes a step further than the smart city; it is not just "smart" but "*super*smart," in that it transcends sectors and strongly emphasizes inter-sector collaboration. One of the greatest technical challenges to this end is to construct a technological development framework under which we can put into operation an information integration architecture and a data platform, which will enable data and information to be integrated between different sectors and will provide a knowledge database linking together the information in different sectors. Thus, the Society 5.0 concept can help spur the technological development necessary for such cross-sector collaboration.

Another task to tackle is to overhaul the traditional model of industry–academia partnership. It will remain important for academic research institutes to steadily advance research projects under commission from or jointly with private companies. But such projects are limited in their capacity to yield systems that can lead society. There are already numerous examples in Western countries of companies and universities collaborating in projects on a common organizational footing. In pursuing Society 5.0, companies and universities should adopt the industry–academia collaboration model, in which they draw on each other's strengths to research a future social vision alongside technological innovation and communicate their findings to a global audience.

On the other hand, Society 5.0 has created an opportunity to develop related technologies such as Big Data analysis and information integration architecture. It has certainly given businesses, universities, and government added impetus to collaboratively develop related technologies, but the opportunities should not be limited to academics in STEM fields, manufacturing businesses, and app developers. The technology underlying Society 5.0 should be broadly defined. Chapter 3 introduced the case of Barcelona, which installed numerous sensors in streets and released the sensory data to the public so that citizens can monitor the data themselves. This approach helped the city address its problems. As this case suggests, a key task in developing the cyber architecture for Society 5.0 is to use IT and Big Data analysis as a means to practically apply ideas for improving citizens' daily lives and living environments.

There remain many facets of our daily lives that are not yet digitized and presented as data. Innovative ideas for digitizing these things and then making use of the data will spark the development of sensor technology and apps for visualizing the data. Given that part of technological development is to unearth the social needs that underlie the technology, we can assume that anyone with a good idea will participate in the process of building Society 5.0's cyber architecture. Moreover, new social systems, such as the sharing economy, have matched individuals' ideas with IT, fleshed out these ideas, and proliferated them in a grassroots manner. Habitat Innovation must be driven by the spontaneous ideas of citizens, who know their habitats well. Likewise, it is citizens who are the end users of the technologies for merging cyberspace and physical space (real world). In these respects, Society 5.0 is a society that facilitates innovation by citizens and for citizens, and that is itself the product of the aggregate of such innovation.

Development of Human Resources and Education

Another prerequisite for Society 5.0 is to ensure that the education system produces experts in the fields where new demand will arise (see Fig. 8.1). When it comes to education, there are two tasks to emphasize.

The first task is to train up experts who use AI to analyze Big Data, as the societal demand for such will grow ever greater in the years ahead. The growing demand for data scientists is attracting attention among genomic Big Data analysts in the fields of medicine and pharmaceutics. With a growing array of IoT-related products, there is now an urgent task to train up experts who can use AI to analyze the Big Data that these products collect. Demand for data scientists is set to soar in fields such as transport (self-driving vehicles), energy (CEMS/BEMS), construction (i-construction), and commerce (e-commerce). Already, universities are struggling to keep up with the societal demand for such human resources. The current crops of university students are not enough to plug the shortfall, so part of the answer lies in recurrent education.

The second task concerns the importance of information literacy in the data-driven society. The general public must gain the literacy to accurately decipher data and information. When you misread data and information, you will tackle a problem in the wrong way, and you might end up using the data or information incorrectly. Suppose, for example, that a local region is experiencing rising crime. The way the local government tackles this problem will depend on how it interprets the crime data. Criminal activity is concentrated in certain hot spots. The local government will use data to tackle the crime problem in either case, but the countermeasures it takes will depend on whether it focuses on the crime hot spots or on the people committing the crimes. It must also consider how releasing this data might impact local communities. Crime data is a classic example of how difficult it can be to interpret data and respond appropriately. Particular care must be taken with open data, as the way the public reacts to the data will impact the local community's future in various ways.

For these reasons, information literacy (how to interpret and use data and information) will become even more crucial in the data-driven society. Educational institutions from elementary school to university will shoulder this task along with

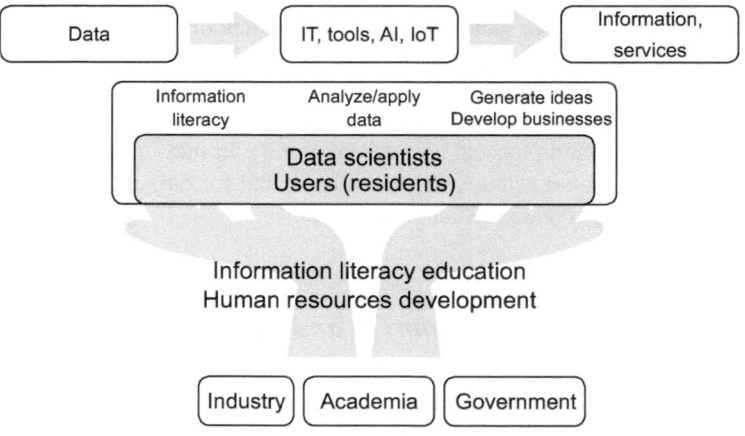

Fig. 8.1 The importance of human resource development and education

companies and local communities, but education in particular will have an essential role to play in helping members of the public gain information literacy. When the public is information literate, the region will become a pioneering example of a true data-driven society, one in which data is used to promote secure and convenient lives in the local community. Educational institutions, businesses, and government each have their role to play in training up the human resources necessary for Society 5.0 and ensuring information literacy.

Promoting Regional Revitalization

The success of Society 5.0 depends on whether national and local governments can assert the political leadership necessary for a strategic shift and institutional reform. There are many different institutional issues that hinder progress in essential tasks such as building an inter-sector information integration architecture and striking a balance between the protection and access to personal information. Moreover, there will be a greater need than ever to ease regulations so as to enable innovation and creation of new business opportunities.

Another issue is that the data of local communities is managed privately and publicly in a decentralized manner, so efforts must be made to consolidate and coordinate the management of such data. To build the inter-sector information integration architecture, government must take a sledgehammer to its vertically compartmentalized systems of data management (see Fig. 8.2). A single set of geographical information is managed and used among assortment of government departments related to construction, roadworks, and sewage systems, so the management of the data must be coordinated. Likewise, data related to transport, welfare, and education must be integrated in such a way that it can be used in other departments. Another matter that cannot be sidestepped is that of personal information protection. Personal data banks and information banks, which hold and use personal information, have burst onto the scene, and they have great potential in the years ahead. Data use is key to Society 5.0. Chapter 3 introduced examples of pioneering local government initiatives in the West and in Japan. These examples illustrate how local governments in rural or provincial areas can benefit when governmental data is opened up to the public, after ensuring human security.

National and local governments must recognize that existing policies will not be enough to balance the resolution of social issues with the demand for pleasant daily lives. They must then reassess the values and principles underlying these policies. Next, they must set new policies and use KPIs to measure their effects. To that end, they must continually collect and analyze data to ensure that policies are grounded in evidence. It should not just be private companies who make use of data. In fact, the Government's Council for Promoting Statistical Reform has advocated evidence-based policymaking (using statistical data as evidence to legitimize and measure the success of policies) (The Council for Promoting Statistical Reform 2017).

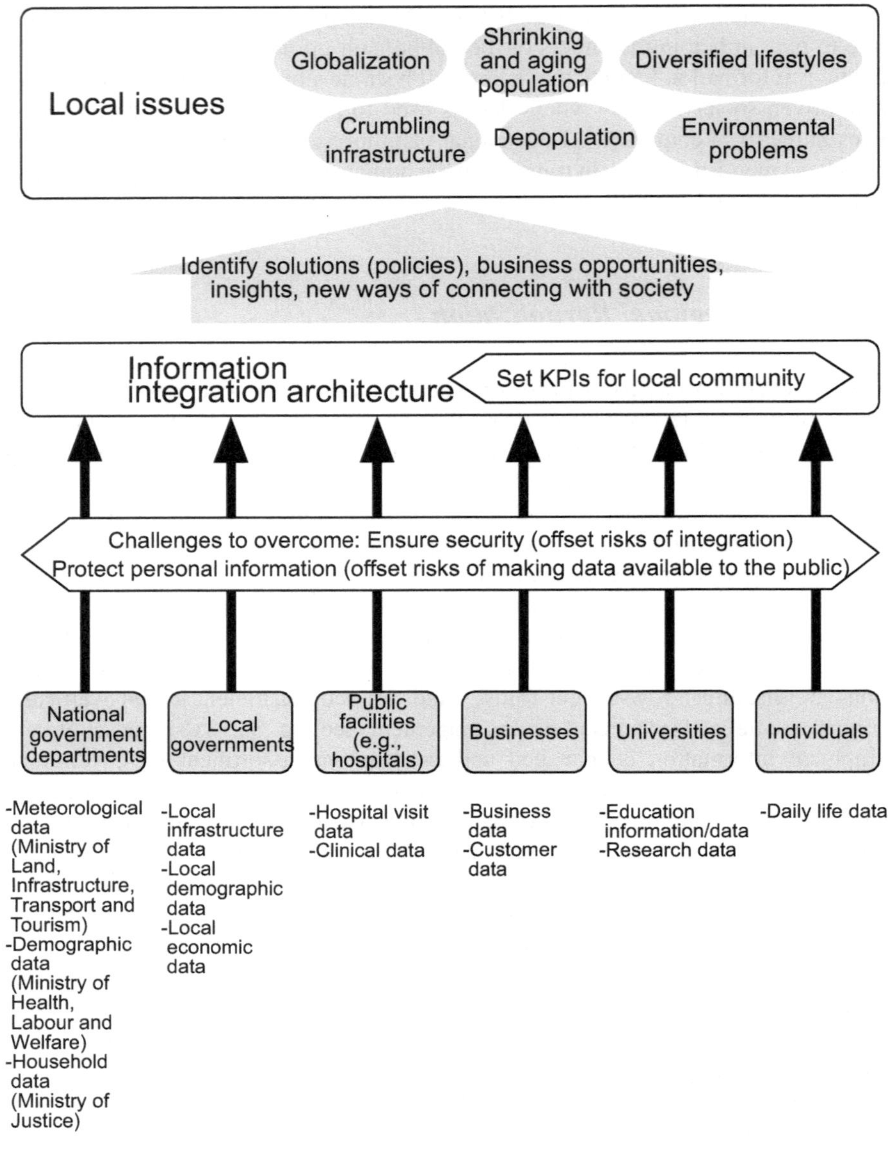

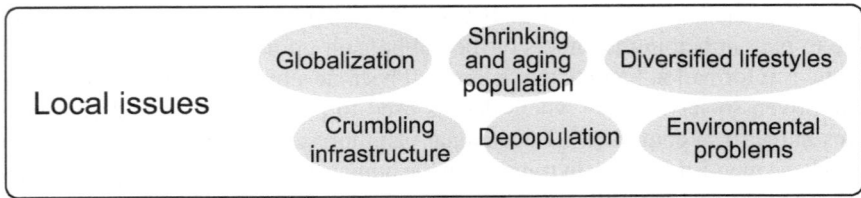

Fig. 8.2 Data integration for regional vitalization, and the relative issuess

In the interests of regional revitalization, there must be an industrial ecosystem. More specifically, local governments and local companies should partake in the vision of the Society 5.0 and promote an IT-based industrial ecosystem to revitalize their local communities. To this end, the national government and local regions must share the common objective of building an ecosystem that can organically link the small businesses and startups that emerge as a result of open data.

To ensure that the Society 5.0 vision of a supersmart society gains traction in provincial regions, there must be regulatory easing in these regions, and government data must be made available there as open data. To ensure that such actions lead to improvements in public services, create business opportunities for new businesses, and encourage universities and companies to collaboratively develop new technologies, there must be an advanced infrastructure that integrates local information networks, and this infrastructure must be used. Local actors must also coordinate industry and academia in such a way as to promote the local area's unique produce and advanced manufacturing as well as a new local service industry.

Society 5.0 as Business Opportunity

Society 5.0 offers a boon to the private sector: the shift from data monopolies to open data will generate new business opportunities. Traditionally, companies have gained profits by monopolizing their customer and marketing data. From now on, companies will create new business opportunities by releasing their datasets as open data (after ensuring human security) and sharing them with others in cyberspace. While paying due attention to personal information protection, companies will publicly release data that they were unable to fully analyze themselves. Bus, railway, and taxi operators will release their people flow data; estate agents will release their data on land and property use; power and gas suppliers will release their data on energy consumption. When all these data are collated and combined, it will surely generate business synergies that the individual companies would otherwise have missed. There will be a great potential for forging new businesses that deliver better services to local communities and users (see Fig. 8.3).

However, there are several challenges to be overcome in proliferating and commercializing the smart city models and initiatives discussed in Chap. 3. For example, in existing smart city projects such as those based on energy management, the aim was to conduct a government-subsidized test bed project and then practically launch the initiative and roll it out in other cities. The challenge in such cases was to make the project commercially viable. The difference with Society 5.0, a society that provides "goods and services that granularly address manifold latent needs," is that the business opportunities extend to members of the public; those with the will to do so can seize these opportunities by using their ideas and insights to forge new IT-based businesses.

In the process of making Society 5.0 a reality, many new business opportunities will arise among universities. The more we progress toward a knowledge-intensive society, the greater opportunities academics have to forge new industries using the body of technology and knowledge accumulated in their research activities. Thus, Society 5.0 expands business opportunities among university students and researchers, and the public at large.

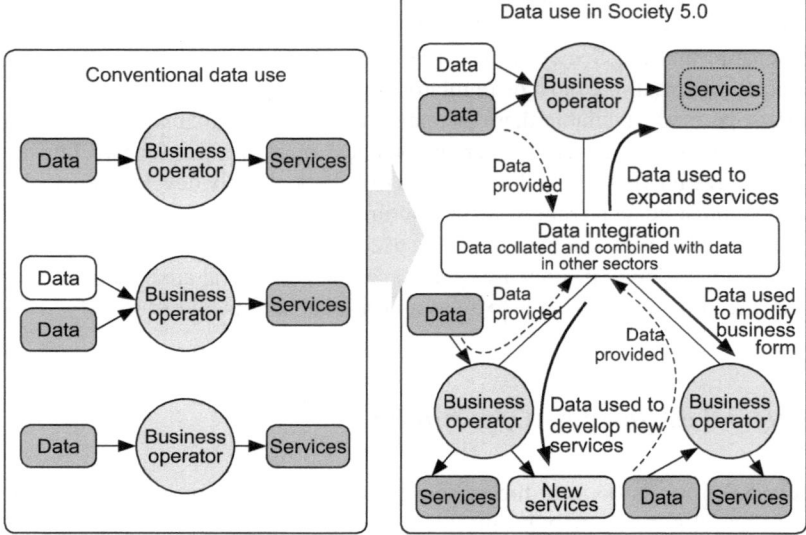

Fig. 8.3 The importance of data integration for creating new businesses

Movement Originating in Japan

Society 5.0 is, in some respects, Japan's global message to the next generation.

Other countries have made headway in applying models in the field of energy management. Broadly defined smart city initiatives are underway in many Western cities. Against this backdrop, Society 5.0 is Japan's homegrown concept for the next generation. There are two aspects to emphasize here. First, Society 5.0 is a vision of a technologically advanced society, one underpinned by Japan's technological prowess. Second, amid the concerns that capitalism will lead to further division, Society 5.0 offers the world a vision of society that is both technologically advanced and people-centric. Whereas Germany's Industrie 4.0 aimed for an IT- and IoT-driven revolution in manufacturing, Society 5.0 includes in its scope the goal of an inclusive society, one that accommodates social diversity and individuals' preferences. Much of the value of Society 5.0 therefore lies in the fact that it presents to the world a future vision that looks beyond technological sophistication, to a people-centric society.

However, if we are to export the idea of a supersmart society that merges cyberspace with physical space (real world), we must tackle a technical and institutional challenge: how to link the cyberspace and information integration architecture across national borders. For the idea of Society 5.0 to gain global traction, the data inside the cyberspace, as well as the cyberspace itself, must be globally standardized. We must work to develop standards regarding the data in sectors such as energy and transport, and standards regarding the process of integrating such data between sectors. In other words, developing standards (such as ISO standards) for sectors related to Society 5.0 is a major ongoing task.

Generally speaking, data should be prepared according to international rules and standards to ensure that it is objective and broad, and that it allows for comparisons with data from other points in time or from other regions. Society 5.0 is critically important for Japan to shift from a problem-stricken to problem-solving developed nation. If Japan succeeds in implementing problem-solving models such as the "vibrant aging society" and the "zero-carbon society," the technology and systems it develops under such models can be exported to other countries and regions in the developing world (such as China and other developing states in Asia and Africa) that are set to face the same kinds of problems. In promoting Society 5.0 as a vision grounded in the country's technological prowess, Japan is providing a forerunner example of a vision-driven society fit for the twenty-first century. Whether this homegrown movement spreads globally will depend on the success of Japan's domestic initiatives.

A Recipe for SDGs

Ever since UN's Sustainable Development Goals, usually abbreviated as SDGs, were ratified in September 2015 at the United Nations Sustainable Development Summit, they have raised awareness of global challenges in Japan and other countries. There are 17 SDGs to guide sustainability policies in countries and regions around the world, and each runs from 2016 to 2030. Under the SDGs, there are 169 targets and 244 indicators (some of which are repeated for different SDGs). The SDGs are ultimately designed to fulfill the UN's pledge that "no one will be left behind." However, it is countries and regions that are responsible for working out how to accomplish the SDGs (United Nations Information Centre 2019; Ministry of Foreign Affairs 2019).

In Japan, companies, universities, and local governments explore ways of contributing to the SDGs, and the Society 5.0 project likewise should accord with the SDG framework.

Society 5.0 is a vision that advocates a technology-driven supersmart society and people-centric society; at the same time, it offers a roadmap for technological progress. Whereas the SDG framework outlines a bottom-up approach to achieving the UN's pledge that "no one will be left behind," the Society 5.0 approach is to facilitate the introduction of cutting-edge technology while also aiming for inclusivity. Insofar as it is a technology-driven vision, Society 5.0 is in large part an attempt to facilitate technological development in each sector. The SDG framework includes cutting-edge technology in its scope too, but it focuses more on solving global challenges such as regional divisions and inequality. As such, the focus of the SDGs naturally falls upon efforts to develop infrastructure (such as sewage works) and public facilities and solve institutional bottlenecks.

SDGs also serve to attract the attention of private companies (which tend to be very profit focused) toward the importance of social contribution and social value creation. They make companies realize that pursuing corporate growth alongside societal development will help the company itself achieve sustainable growth. In this way, they are effective in inducing companies to develop approaches and targets to such an end. The 17 SDGs cover a comprehensive set of themes, such that any company in any industry can find a way to contribute to at least one of them. This design means that the SDGs can easily be incorporated into the action plans of national and local governments and other core components of society such as companies and universities.

In the case of the SDGs, the main outputs are the degree to which the goals have been accomplished and how they are being tackled. In the case of Society 5.0, the aim is to use technology to balance economic advancement with the resolution of social problems, and in the course of this balancing process, technologies and the systems for pioneering and rolling out these technologies emerge. The outputs of Society 5.0, then, are the innovative technologies and systems as well as the resulting problem-solving models, which describe how problems were solved by introducing said technologies. These models can then serve as recipes on how to accomplish the SDGs. The outputs of Society 5.0—vision-driven technologies and systems—will emerge in society one after the other, providing ways to accelerate progress in the SDGs.

In summary, Society 5.0 has multifaceted significance. As concept of a technology-driven society that purports to be supersmart and people-centric, Society 5.0 does not just provide a vision to guide Japan's science and technology strategy; its relevance extends to the political and economic spheres, and it offers abundant hints on how to forge a future society.

The spread of IT applications is taking us steadily closer toward the supersmart society. But we still have no guarantee that the supersmart society would become a people-centric society as the fledgling vision suggests. We harbor the fear that future advances in IT and AI may, as has sometimes been the case, lead to a more inhumane society.

Thus, when it comes to Society 5.0, perhaps the most important thing is to keep in mind how technological innovation should lead society in a better direction, and to ensure that the principle of the people-centric society occupies the hearts and minds of the actors and organizations involved in technology development and community development, as well as the hearts and minds of the engineers and of each and every member of the public.

References

Cabinet Office (The Commission on Measuring Well-being) (2011) Measuring national well-being-proposed well-being indicators. https://www5.cao.go.jp/keizai2/koufukudo/pdf/koufu-kudosian_english.pdf. Accessed 5 Jun 2019

Ministry of Foreign Affairs (2019) Japan SDGs Action Platform. https://www.mofa.go.jp/policy/oda/sdgs/index.html. Accessed 5 Jun 2019

The Council for Promoting Statistical Reform (2017) Tōkei kaikaku suishin kaigi saishū torimatome [Final Report of the Council for Promoting Statistical Reform]. https://www.kantei.go.jp/jp/singi/toukeikaikaku/pdf/saishu_honbun.pdf. Accessed 5 Jun 2019

United Nations Information Centre (2019) Sustainable Development Goal. https://www.un.org/sustainabledevelopment/. Accessed 5 Jun 2019

Correction to: Society 5.0

Correction to:
Society 5.0, **https://doi.org/10.1007/978-981-15-2989-4**

The original version of the book was revised with the correct Institutional Editor's name as "Hitachi-UTokyo Laboratory (H-UTokyo Lab.)."

This book was inadvertently published with the incorrect license type CC BY 4.0. The Open Access License has been amended throughout the book to the correct license type CC-BY- NC-ND.

The updated online version of the book can be found at
https://doi.org/10.1007/978-981-15-2989-4

C1
Society 5.0, https://doi.org/10.1007/978-981-15-2989-4_9

Afterword

Atsushi Deguchi and Shigetoshi Sameshima

In June 2016, H-UTokyo Lab. set the goal of making Society 5.0 a reality. At the time, few people had even heard of Society 5.0. This was only natural given that the concept was still very new. The Cabinet's 5th Science and Technology Basic Plan, which first set out Society 5.0 as a key concept, had been released only some months earlier, in January. Back in the early days of H-UTokyo Lab., we would mention the term "Society 5.0" during discussions with experts, and the reaction was usually one of bewilderment. We have come a long way since those times. The project to produce this book came to fruition in spring 2018. Although more people in Japan have now heard of Society 5.0, they often have only a tenuous grasp of the term. The term had already been discussed in a range of different settings, but this publication was intended to promote a shared understanding of Society 5.0 by clarifying the concepts and thinking behind it.

However, even the authors of this book initially had different ideas and interpretations about the key concepts. H-UTokyo Lab. therefore organized a series of symposia with the authors to ensure that we were all on the same page and to further explore the relevant concepts and the way to approach technological developments. The symposia were attended by the teaching staff of the University of Tokyo from the natural and social sciences. These individuals provided invaluable ideas and insights, which have been incorporated into this book.

The research on how to make Society 5.0 a reality must focus on two components together: policy proposals and technological development. Accordingly, H-UTokyo Lab., under its industry-academia collaboration model, pursues both

A. Deguchi
Department of Socio-Cultural Environmental Studies, Graduate School of Frontier Sciences,
The University of Tokyo, Tokyo, Japan
e-mail: deguchi@edu.k.u-tokyo.ac.jp

S. Sameshima
Center for Technology Innovation, Research & Development Group, Hitachi, Ltd.,
Tokyo, Japan
e-mail: shigetoshi.sameshima.uc@hitachi.com

Society 5.0, https://doi.org/10.1007/978-981-15-2989-4

components as a single set with a team that comprises not only Hitachi's technicians and engineers but also academics from a wide range of fields, including engineering, economics, and psychology. Some of the research outcomes are included in this book. The industry-academia collaboration model extends the hand of collaboration even further to include government and industry. With government and industry actors on board, the project will gain greater clout and start making a broader social impact. Industry-academia collaboration has great potential as a model for businesses and universities to combine forces, link up organizationally, and work to resolve social problems and produce technological innovation.

Society 5.0 was proposed partly as a response to the challenges Japan faces as a developed economy. For Japan to shift from a problem-stricken to a problem-solving developed country, each layer of society—central government, cities, and local communities—must overhaul the systems and practices they created during the country's high-economic-growth period. Practical efforts in this direction are already underway. Against this backdrop, Society 5.0, in outlining a new tech-driven society, constitutes a vision for combining the resolution of social problems with Japan's next phase of economic growth. The proliferation of IT in the so-called digital revolution is producing seismic shifts in industry and society. Japan must embrace the change and create a positive cycle that produces new industries. As the concept of VUCA describes, we live in "Volatile, Uncertain, Complex, and Ambiguous" times, which require us to establish and share goals on how to deal with the situations.

This book has discussed the Society 5.0 approach for transforming cities and the outlook for such an approach. The goal of transforming cities is to enhance resident value. Until now, city planners have been responsible for designing and providing habitat value. In the future, however, residents will be the ones who create this value. Social media and digitalization in manufacturing are creating a paradigm shift whereby end users are engaging in industrial processes. The primary agents of the new society are individual citizens. Likewise, when it comes to the transformation of cities, individual citizens will play the leading role as innovators, creating desired urban environments afresh or overhauling existing ones. Such citizen-led innovation is key to Society 5.0.

However, citizens cannot do the work on their own. Each sector of society must play its part in helping Japan make Society 5.0 a success in the face of a competitive world, where countries jostle to get ahead. Industry, for its part, must make clear its commitment to improving existing urban environments and create new business opportunities by releasing and promoting the use of data that could prove effective for habitat innovation. Government, for its part, must promote a model of community development that is grounded in the unique features of local communities. It must do so by overhauling and fine-tuning legislation and by creating systems that enable the safe use of data. As for academia, it has a role to play in creating the new social systems for a post-digitalization world. Drawing on the insights of the humanities, universities must spearhead urban reforms while anticipating future challenges.

Japan is in the midst of the new digital revolution. Industry, academia, and government must each play their part and work together so that Japan can get ahead of the competition and so that we can establish models based on the concept of the people-centric society.

As the new digital revolution, which some call data capitalism, takes the world by storm, Society 5.0 presents to the international community Japan's vision of a future digital society. We hope this book helps accelerate the industry-academia-government collaboration necessary to achieve this vision.